Fishing with Electricity

The Author

Dr K P Biswas, M.Sc., Ph. D., D.F.Sc. (Bombay), E.F. (West Germany); F.Z.S., F.A.B.S. (Kolkata) initiated the work on Electro-physiology and Electric Fishing, first in India, in 1961 at Odisha. He developed an electric seine to catch fishes from large and deep waters and reservoirs with the aid of portable AC generator and successfully caught bottom and burrow dwelling fishes, which otherwise, are difficult to catch with any other fishing gears. He determined electro-physiological constants of several fishes of freshwater and marine habitat. In 2015, the author has developed a DC electric shocker to catch fishes without any net. In 2017 and 2018, he guided eleven M.Sc. students of Calcutta University for their thesis work, "On the reactions of twenty Indian fish species in DC and pulsed DC electric field.

Fishing with Electricity

– Author –

K P Biswas

2019

Daya Publishing House®

A Division of

Astral International Pvt. Ltd.

New Delhi – 110 002

Published by   : **Daya Publishing House**®
                          A Division of
                          Astral International Pvt. Ltd.
                          – ISO 9001:2015 Certified Company –
                          4736/23, Ansari Road, Darya Ganj
                          New Delhi-110 002
                          Ph. 011-43549197, 23278134
                          E-mail: info@astralint.com

Digitally Printed at : **Replika Press Pvt. Ltd.**

Dedicated to

Mrs. *Manju Biswas* for Encouragement

Acknowledgement

The author deeply acknowledge the help of Dr N A Talwar for sending the manuscript on line to the publisher.

Dr K P Biswas

Preface

Fishing with electricity first started with the patent granted to Mr. Isham Baggs, living in Islington, London in 1863. Following Isham Bagg's work, electric fishing developed in different regions in different environments. Notable among them is the work of Holzer (1932) in Germany. In the early 20th century Japanese fishermen were using induction coil to drive eels out of their burrows into a net. Larssen in 1912 took a patent for the use of electricity to catch a variety of aquatic organisms, from fish to seals.

Scientists use electrofishing for fish population survey in aquatic environment, mainly in the rivers and streams to monitor the size of population and determine the species in a community.

The basic principle involves in creating an electrical field in the water to excite the muscular system through peripheral nerves for involuntary movement of fish towards the electrode and get stunned. When the fish are stunned, they often float near the surface of the water and can be removed from the electrical field.

When a fish swims into a weak electrical field, it may not be affected at all. A threshold electric field should be created in the water in order to affect the fish.

When the electrical strength in the water is sufficient to allow transport of charged ions across the nerve cells of fish body, the fish's muscle will undergo involuntary contraction. The contracts will result to increased exercise of muscle and build up lactic acid in blood stream and there will be a cramp in the fish muscle. When the fish cramps up, it floats to the water surface and can be caught it effectively. Usually 5 to 10 seconds are required to stun the fish.

Once the fish is removed from electrical field and kept in aerated water, the aeration and cool temperature will recover the fish after paying the oxygen debt. The lactate in the blood stream and cramping must be treated with rest and good oxygen supply. The recovery time of fish is generally a minimum of four hours, but may be extended up to twelve hours.

The author, in India, had developed an "electric seine net" in 1962 after conducting laboratory experiments with Indian fish species. This electrically charged net was successfully used at Dhauli, Odisha for catching fishes from burrows and bottom dwellers along with columner fishes from earthen ponds of various sizes and depths.

He also developed a DC fish catcher producing electric field of 110 volts in the water. The catcher has been successfully tested with suitable electrode configuration and demonstrated before an ICAR peer review team on 31st August, 2016 in the Fisheries Faculty of West Bengal University of Fisheries and Animal Sciences. The team leader suggested the author to make the technique aware to the people of our country.

The suggestion initiated the author to prepare the manuscript of the " Fishing with Electricity ".

Dr K P Biswas

Contents

Acknowledgement   *vii*

Preface   *ix*

Introduction   *xv*

1. **What is Meant by Electrofishing ?**   1

2. **Principles Behind Electric Fishing**   5

    *Basic Electricity — Electrical Current Types — Alternating Current (AC) —
    Direct Current (DC) — Pulsed Direct Current (PDC) — Electrophysiology
    — Circuit Principles — Example — Example — Waveforms — Fish Reaction
    and Sensitivity — Fundamental Electrofishing System — Maximum Power
    Transfer — Check the Setting of This Figure — Checking the Electrofishing
    System using an Oscilloscope — A Typical Cathode Ray Oscilloscope — Setting
    up the Oscilloscope — Detailed Method to Obtain a Trace — Further Setting
    up Procedure — Practical Experiences — Electrical Field Theory — Electrode
    Configuration — Safety — Standardization — Testing of Electrofishing
    Equipment in the Field*

3. **History of Electrical Fishing**   35

4. **Electrical Fishing Technique Adopted in Different Regions**   39

    *Passage of Current through Water — Development of Electrodes — Operation
    of the Shocker — Electric Seine — Limitations of the Method — Equipment —
    Operation — Tests of Efficiency*

5. **Electrofishing Boat**   47

Problem Areas — General Design — Components of an Electrofishing Systems — Boat and Mechanical Components — Electrical System — Main Generator — Auxiliary Generator and Battery — Meters — Disconnection and Overload Protection — Voltage Control — Interconnection System — Safety System — Electrode Systems — Electrode Requirements — Basic Electrode Phenomena — Water Conductivity Effect — Electrode Design in Alternating Current (AC) — Electrode Design in Direct Current (DC) — Instructions for Field Study — AC Operation — DC Operation — PDC Operation — Operational Methods — Electrofishing Boat — Mechanical Configuration — Electrical System — Transformer Unit — Pulser — Electrodes — Cathode Array — Anode Array — Electrical Characteristics — Results of Electrofishing Experiments — Unsuccessful Experiments — Guidelines for Operation — Summary of the Field Tests with the Electrofishing Experimental Boat — Electrofishing Boat using AC Current — Mechanical Configuration — Electrical System — Engine-Generator and Generator Control Unit — Transformer Unit — Electrode System — Cylindrical Electrode Arrays — Electrical Characteristics — Electrofishing Results — Results of AC Electrofishing

6. **Fishing with Electricity in other Countries**   79

Direct Current Electrofishing in Streams of Canada — Wolf Power Plant — Pulse-Modulated DC Electric Fishing Machine to Capture Eels in New Zealand — Design of the Machine – Physiological Basis — Electrofishing Shocker Developed by the Bureau of Commercial Fisheries, USA in 1966 — Fish Reactions to Increasing Current Strength — Techniques of Fishing with Electricity — Considerations for Electrofishing

7. **Electrical Fishing in India**   89

Development of Electric Fish Catcher

8. **Electrical Fishing in Sea Water**   97

European Electrofishing Gears — "Pulse" Trawling for Flat Fish — Hovercran Trawl for Shrimp — Capture Mechanisms — Response by Fish — Response by Shell Fish — Environmental Effects — Effect on Cod — Effects on other Fish Species — Effects on Benthic Invertebrates — Trawl Path Mortality — Chemical Effects — Comparisons between Gears — Comparative Trials — Surveys of Commercial Fishing — Fisheries Management Implications — Pulse Trawling Research — Control of Electric Field Characteristics — Hovercran Shrimp Trawling — Experiments with Pelagic Fishes for Design Criteria for Electrical Fish Harvesting Systems — Test Equipment and Procedure — Test Procedure — Results Obtained — Captured Fish — Wild Fish — 120-kVA Pulse Generator Design — Netless Fish Harvesting Mode — Mid-Water Trawling Mode

9. **System Components – Electric Fishing Machine**   119

Electro Fishing Power Supply – Design Consideration — System Operation — Safety Measures — Engine-generator Power Source — Transformer Panel — Transformer Winding Specifications — Power Control Panel — Timing Circuit — Hand-Crank Magnetos as Fish Stunner — Simple Electric Fish Shocker Circuit — Electric Fishing Gear Types — Gasoline Engine Generators — Battery Operated Gear — Electrodes for Freshwater Electrofishing — Ranges of Efficiency — Impulse Generators for Trawl Fishing — Use of Generator in Electrical Fishing Gear — Electrodes for Sea Fishing Gear

10. **Factors Affecting the Efficiency of Electrical Fishing Gear**   137

11. **Electrofishing Induced Injury and Harmful Effects**   141

Physiological Effects — Implications on Non-target Species — Factors Affecting Injuries and Mortalities — Impacts on Reproduction, Embryos, and Larvae

12. **Hazards of Electrofishing: Precautions and Safety Measures**   151

13. **Electrofishing Policy and Code of Practice**   155

Hazards of Electrofishing — Operators and Crew Training and Certification — Construction Standards of Equipment — General — Backpacks — Shore-Based Units — Boat-Mounted Units (min. 3.5 m) — Suppliers — Maintenance Requirements — Personal Insulation and Safety — Operations Practices — Public Safety — Weather and Site Limitations — Working Procedures — Care of Fish — Care of other Fauna — Transfer of Biological Material — Accidents and Emergency Procedures — Electrofishing Apparatus Safety Features Checklists — Equipment Type – Shore-based — Equipment Type – Boat-mounted — Equipment Type – Back-pack

14. **Electrofishing Terminology**   167

General — Electric Field Characteristics — Behavior and Physiology

References   173

Index   181

Introduction

Fishing with electricity, the use of electric current in water to stun fish for capture, is a powerful tool for sampling technique for nearly a century. Scientists working with fish and other aquatic organisms are using this technique to determine the status of a population in lentic and lotic waters, especially in case of endangered or threatened ones that are impacted by anglers or fishers.

Exclusively for fishing, electricity was first used in 1863, when a British patent was graned to Mr. Isham Bagg living in Islington, London. Following his work, fishing with electricity developed in different regions in different environments. In the early 20th century Japanese fishermen were using induction coil to drive eels out of their burrows. Larssen in 1912 took a patent for the use of electricity to catch a variety of organisms from fish to seals.

But the technique involves a very dynamic complex and poorly understood mix of physics, physiology and fish behavior.

When a fish swims into a weak electric field, it may not be affected at all. There is a threshold of electrical charge that must be emitted into the water in order to affect the fish. To be effective, the electric field in the water must be sufficiently strong at appropriate distances from the electrodes to elicit desired responses by targeted fish. When the electrical charge in the water is sufficient to allow transport of the charge across the nerve cells in the body, then the fish's muscles will undergo involuntary contraction. The contractions will lead to increased exercise of the muscle and a build up of lactate in the blood stream.

The size, shape and nature of electrical field are defined by the distribution and pattern of electrical intensity which is determined by the peak electrical potential (voltage difference), type and intensity of current and waveform generated between and around electrodes; the position, size and shape of the electrodes, the conductivity of the water and bounding and surrounding media; and the size and dimension of water bodies.

The cumulative results of many years of individual and piecemeal investigations is now known to be the responses of fish to electrical fields. Many questions remained and the interpretation of some results is difficult to understand.

In mid-fifties, many researchers put emphasis on the current density of the field to elicit different responses of fish; which together with voltage gradient between the body extremities, referred to, as " body voltage" (voltage difference between head to tail) was thought to be the cause of involuntary taxis of the fish. Orientation of fish's body axis with relation to lines of current conduction was reported to initiate perception and movement. Fishes in a homogeneous field at right angle to the equi-potential are subjected to the highest potential difference and the fish was found to move eratically or to tap off minimum potential difference between their body extremities by placing them parallel to equipotential lines, a condition, termed as oscillotaxis.

A new hypothesis has now been advanced regarding "power transfer" to fish and the epileptic nature of their responses to electric fields. To better understand and explain the interaction between fish and electric fields, electrofishing has been treated as a power-related phenomenon. According to this " power-transfer theory for electrofishing ", the relationship between electrical power in the water and in the fish is a function of the ratio of water conductivity to the effective conductivity of the fish (Kolz and Reynelds, 1989; Kolz et al., 1998). More recently, it has been suggested that the observed reactions of fishes to electric field, including twitches (in the zone of perception or reactive detection) taxis, narcosis and tetanus, are essentially aspects of the same phases of epilepsy (automatism, petit mal and grand mal) that are observed in humans and other animals subjected to electroconvulsive therapy (Sharber et al., 1994, 1995; Sharber and Black, 1999). Most of the recently accepted or proposed concepts for explaining or better understanding the reaction of fish to electric fields and the mechanism involved need to be further explained, validated, refined and integrated for the advancement of science and technology of electrofishing.

As regards the harmful effects, stress, injuries and sometimes mortalities among captured fish are unavoidable consequences of electrofishing and other fish catching techniques. Among the most effective gear and techniques available for catching fish, biologists usually select those known to be least harmful.

The electric field quickly weakens with distance from the electrode and with depth. So many fish outside the influence zone are never affected. Some fish that have been stunned may even build up a tolerance for the electric field. Many species also have evolutionary adaptations, such as a keen lateral sensory system and eye sight, that help them to avoid the influence zone. Many other species, such as carps have thick scales that protect them from conduction of electric charges in the water. Small fish such as juveniles and minnows are so small in size and body area, that the electrical field does not affect them.

Generally fish recover almost immediately after they are stunned and removed from the electrical field. They become oriented in the upright position and begin to swim normally within 1-2 minutes. To fully recover from the stress of electrofishing, it can take 4-12 hours, which depends on the amount of lactate in the blood or the level of stress the fish experiences and habitat conditions. It may also take longer time for the fish to recover when the dissolved oxygen in the water is low.

Whether the fish dies or not, because of electrofishing, depends on the operator generating the electric field in the water and the handling techniques that follow. Most fish survive through the experience, but delayed mortality following the electrofishing is not well studied.

Fish may die immediately if they are shocked too intensely. This is evident because some of the skin tissue begins to turn black. This type of mortality is one of the easiest to avoid, because it is easy to detect and easy to correct.

Fish may experience some delayed mortality because of lactic acidosis, which is build up and persistence of lactate in the blood stream. If the lactate is not removed by sufficient respiration of oxygen or if the fish can not adapt to high levels of lactate, then it will die.

Prior to the late 1980s, electrofishing has been considered not only the most effective, but also the least harmful means to capture fish, particularly moderate to large-sized ones.

Biological investigations of electrofished fish revealed large percentages of fish with electrofishing injuries, especially salmonid species (Holmes *et al.*, 1990; Meyer and Miller, 1991; Fredenberg, 1992; Newman, 1992; Mc Michael, 1993; Hollender and Carline, 1994). These injuries are not externally obvious, even if severe ones. Factors considered affecting electrofishing injuries, though sparse, include, type of current, field intensity, duration of exposure, orientation of fish to lines of current, and for alternating current (AC) and pulsed direct current (PDC), waveform characteristics, such as, shape, pulse frequency and pulse width. Other factors considered were fish species, size and condition.

Among three types of electrofishing currents, AC is considered most harmful, DC (constant direct current) least and PDC (pulsed direct current) usually somewhere between depending on the frequency and complexity of pulses. The extent of mortality or injury caused by each of these currents varies considerably with the way of their use, other electrical parameters, biological factors and environmental conditions.

Similar to the concentration of most chemical substances and physical parameters affecting living organisms, field intensity and duration of exposure are the primary factors affecting physiological stress and mortality in fish subjected to electrofishing currents. Beyond, it is not the field intensity itself, but the magnitude of voltage differential generated across the fish (head to tail voltage) or specific nerves or tissues that causes electrofishing mortalities. Differential voltage is a function of both field intensity and orientation of fish relative to the lines of current.

The electrofishing has advanced considerably in equipment technology, safety and sampling design; however, it has remained static in our understanding of electrophysiology, the response of fish to electric currents and factors affecting the efficiency of electric fishing.

A questionnaire to identify the present use of electric fishing was sent to 25 European Inland Fisheries Advisory Commission (EIFAC) national correspondents. Replies from 30 agencies in 10European countries were received. Classical wading and classical boat fishing(two hand-held anodes) on large water bodies are the main electric fishing activities in Europe. In the United States, boat fishing with the boom-mounted arrays is most common (Lazauski and Malvestuto, 1984).

Dr. K.P. Biswas

Chapter 1

What is Meant by Electrofishing ?

Electrofishing can be defined as a fish sampling technique using electric currents and electric fields to control fish movement and/or immobilize fish, allowing the easy capture of fish. It is a technique whereby electrical energy is put into the water and fish, intercepting this energy, are drawn toward the probes and incapacitated in such a way that they can be captured with nets. The movement of fish toward the source of electricity is called galvanotaxis (uncontrolled involuntary muscular convulsion that results in the fish swimming toward the anode) and is believed to be a result of direct stimulation of the central and autonomic nervous systems which control the fish's voluntary and involuntary reactions.

The effectiveness of electrofishing is influenced by a variety of biological, technical, logistical, and environmental factors. The catch is often selectively biased as to fish size and species composition. When using pulsed DC for fishing, the pulse rate and the intensity of the electric field strongly influence the size and nature of the catch. The conductivity of the water, which is determined by the concentration in the water of charge carriers (ions), influences the shape and extent of the electric field in the water and thus affects the field's ability to induce capture-prone behavior in the fish. Fishing with electricity or electrofishing is one of the most common method used in freshwater fisheries research. Many research organizations and agencies, in many parts of the world employs electrofishing as the primary method for assessing fish communities in stream monitoring program. Though electrofishing is a common sampling method, many biologists and field technicians are not conversant about electrofishing theory, equipment configuration, implementing the sampling design and data interpretation. They often encounter several difficulties and uncertainities in optimizing electrofishing configuration, implementing the sampling protocol and interpreting the results.

Peer-viewed literatures, during examination also reveal that most did not provide enough information on how electrofishing was conducted to evaluate electrofishing efficiency and data quality. Since electrofishing is thought partly science and partly art, it may be possible to replicate results but impossible to achieve full comparability between studies performed by different workers using different equipments in different conditions. There is room, however, to expend our understanding of electrofishing principles to improve our sampling designs, and to communicate methods and interpretation of results more clearly and completely to other fishery biologists and environmental managers.

Scientists use electrofishing to determine the presence of fishes and monitor the size of populations and the species in a community. Whether using a boat or a backpack electroshocker, the basic principle involves creating an electrical field in the water to immobilize fish. When fish are stunned, they often float near the surface of the water and can be removed from the electrical field. Some scientists monitor the status of fish populations, especially endangered or threatened ones, or ones that are impacted by anglers or fishers. Scientists working with fish or other aquatic organisms have various tools for ascertaining the status of a fish population, especially endangered or threatened ones, or ones that are impacted by anglers or fishers. Electrofishing from a boat is a powerful tool to survey tidal habitats with silty bottom that are not waded easily.Electrofishing may also be less harmful to individual fishes than some other methods, such as, gill netting.

When a fish is subjected to a weak electrical field, it may not be affected at all. There is a threshold of electric charge that must be created into the water to affect the fish.

When the electrical charge in the water is sufficient to allow transport of the charges across the nerve cells of the fish body, then the fish's muscle will undergo involuntary contraction. The contractions will lead to increased exercise of the muscle and build up of lactate in the blood stream. The process is very similar to what happens to the muscles of a runner or swimmer who exerts a lot of exercise. The runner or swimmer may eventually get a cramp in the muscle and can not move it effectively. When the fish cramps up, it floats to the surface and can be removed from the electrical field. The process to stun a fish usually take 5-10 seconds.

Once the fish is removed from the electric field, the fish is kept in a live well with good aeration. The aeration and cool temperatures are essential in the live well, because the fish has an oxygen debt that needs to be paid. Like a runner or swimmer, the lactate in the blood stream and cramping must be treated with rest and good oxygen supply. While this period of recovery may take a short time in humans, the recovery time for fish is generally, a minimum of four hours but can range up to twelve hours.

Not all the fish are likely to be stunned by electrofishing. Some species are easily stunned by the electric field; while the others are not. The electric field quickly weakens with distance from the electrode and with depth, so many fish in the area are never affected. Some fish that have been stunned may even develop a tolerance for electric field. Many species also have evolutionary adaptations that help them avoid the electrodes, such as, a keen lateral sensory system and eyesight. The black bass is able to see the approaching electrodes and swim out of the way of the electric field. Species having thick scales (carps) protect them from the electric field. Electrical field does not affect the smaller fishes due to their small size.

Generally, fish recover almost immediately after they are narcosed in the electric field. They orient themselves in the upright position and begin to swim normally within 1-2 minutes. To fully recover from the electrofishing shocks, it can take 4 to 12 hours, which depends on the amount of lactate formed in the blood or the level of stress the fish experiences and habitat conditions. When the dissolved oxygen in the water is low, fish take longer time to recover.

The death of fish depends on the field strength and the time of exposure the fish is subjected to; which again depends on the person generating electric field and the handlings techniques that follow.

Fish may die immediately if they are shocked intensively, clearly evident by turning some of the skin tissue to black.

Fish may be subjected to delayed mortality, because of lactic acidosis, which is the build up and persistence of lactate in the blood stream. If the lactate is not removed by sufficient respiration of oxygen or if the fish can not adapt to high levels of lactate, then it will die.

Chapter 2

Principles Behind Electric Fishing

The function of an electrofishing system is to produce an adequate electrical stimulus in the fish body near the electrodes to permit easy capture by netting. In an alternating current (AC) system the electrical stimulus only immobilizes fish (electronarcosis or electrotetanus) requiring the net handlers to collect fish from considerable depths or distances from the electrodes. Often fish are immobilized at depths beyond reach of the dip nets and are not susceptable to capture at all. In a direct current (DC) or pulsed direct current system (PDS), fish near the anode exhibit forced swimming (electrotaxis) toward the anode and are more easily netted. It is well established that fish are much more susceptible to AC than to DC and hence the radius of action of an AC electrofishing system operating at a particular voltage is much greater than DC system at the same potential difference. Pulsed DC is more effective than DC and to some extent combines the desirable forced swimming response of DC with greater radius of action associated with AC.

The primary function of the electrofishing system is to establish an electric current in the water near the fish. A portion of this current (depending upon the ratio of fish and water conductivity) will pass through the fish and if of sufficient magnitude will elicit the desired response. The important point is that the current is the variable of concern. Electrode voltage, spacing, size *etc.* are important parameters only to the extent that they influence the current.

Based on this principle, for any electrofishing syatem there exists some minimum value of current per electrode which will affect fish out to some specified distance from the electrode. The minimum current will depend upon electrode shape, nature of current (DC, AC or pulsed DC), desired range of action, fish species, water temperature, water conductivity and many other less important factors. Of these factors, only electrode shape and current type are under control of the designer. All other factors are external variables which must be compensated for by variations in the operation system.

Range limitation, the distance at which fish are affected can be too small to produce useful catching results. This is often caused by improper electrode arrangements and insufficient current and power.

Low water conductivity makes it very difficult to attain sufficient currents to produce useful electrofishing responses. Extreme conductivities in Appalachian mountain streams was resolved by adding salt to the water to improve electrofishing (Parker, 1958).Extremely high water conductivities call for currents too large to be supplied by portable equipment without special electrical control methods. For very high conductivity, DC is ineffective (Vincent, 1971) and pulsed DC or AC may be more effective. Other important factors are voltage and electrode size.

Poor water clarity and excessive vegetation restrict visibility and reduce the value of the immobilizing capabilities of the AC electrofishing. Pulsed DC and DC offer potential solutions to this problem.Fishes which are immobilized at depths exceeding 0.9-1.2 m (3-4 feet) are very difficult to capture by ordinarry netting procedures. Pulsed DC and DC help to overcome this problem.

High conductivity bottom materials tend to "short circuit" the current out of the water into the bottom material reducing the electrofishing effectiveness and sometimes overloading the power source. Except to use small electrodes, near the surface to avoid direct contact between electrodes and the bottom, there is no remedy of this situation. The conductivity of water and water temperatures are more or less a straight line relationship, that is, as the temperature of the water increases, the conductivity of water increases at an equivalent rate (Sigler, 1969). Theoretically, success of electrofishing should increase with temperature, but workers in the field noted differently. Smith and Elson (1950) believed that salmon parr exhibited the best response below 25 degree Celsius and suckers at less than 20 degree Celsius. Webster, Forney, Gibbs, Severns and Van Woert (1955) had greater success in shocking brown trout at 7.8 degree Celsius than at 16.7 degree Celsius, when both AC and DC were used. The response of brook trout to DC current is low in cold water, but increases with temperature to 10 degree Celsius. As the temperature increases further, the response decreases (Elson, 1942). Most salmonids are more easily captured by electrofishing when the water temperatures are low, 0 to 10 degree Celsius (Vincent, 1971).

Fish mortality caused by AC electrofishing are higher than those caused by DC or pulsed DC (Tayler *et al.*, 1957) and the gross physical damage from AC can be severe (Hauck, 1949). Harmful effects from pulsed DC are usually a result of excessive exposure or intense electrical fields (Pugh, 1962). It can be demonstrated that fish can be killed with electricity in their natural environment, as well as, under laboratory conditions (Pratt, 1954; Godfrey, 1956).

With electrofishing boats, mortality is usually associated with regions of excessive high current density close to the electrodes. Probable solutions involve electrode designs which avoid such regions of high current density by using larger electrodes and lower voltages. No injury of fish has been observed with DC or Pulsed DC.

Individual variation is notable among fish even though they are of the same species and similar lengths. The laboratory of Haskell *et al.* (1954) on brown trout demonstrated this variability.

The larger the individual of a species, the more sensitive it is to a given electric shock (Mc Millan, 1928; McLain and Nielsen, 1953; Taylor *et al.*, 1957; Biswas, 1976). Fish absorb power as a function of body surface area and particularly length (Holzer, 1931). Response to shocks were further reduced in case of smaller fishes. All these factors contribute to capturing a greater proportion of large fish than are actually present in the population.

Some species are quite difficult to capture by electrofishing. The most notable are northern pike and muskellunge. It appears likely that this is caused by the strong swimming ability and natural tendency to escape danger by rapid swimming coupled with high sensitivity to electric fields.

Greater differences in response have been observed between fish species when pulsed DC was used. This difference give a scope for some selectivity in electrical fishing. Optimal wave forms and pulse frequency will further produce selective and efficient electrofishing for various species.

The knowledge required includes an understanding of both basic electrical principles and the properties of water.

Safe and effective electrofishing requires an understanding of the basic principles of electrical circuit and field theory, which is the study of electrical energy dissipated into a three dimensional medium, the water in this case. Most people who electrofish spend much of their time focusing on circuits, boat wiring, electrical units, metering and other hardware aspects of electrofishing. However, one must understand field theory to apply electrofishing principles correctly. Based on what happens in sampling situations, electrofishing mythologies emerge that may not have anything to do with electrical principles.

Basic Electricity

All matter consists of charged particles that attract or repel each other because of the positive or negative charges they bear. Electricity is the form of energy that results from this attraction or repelling of particles. Electricity can be defined as the force that moves electrons (Smith, 1989). A circuit is a closed path along which an electric charge moves. The rate of flow or intensity that moves the charge is the current, which is measured in amperes. The electromotive force that moves the current is voltage and is measured in volts. Voltage may also be defined as the potential force available to move electrons through the circuit. The restriction of electron flow in the circuit is resistance and is measured in ohms. Electrical power is the rate at which electrical work is done and is measured in watts. One watt of power results when a current of one ampere flows through a resistance of one ohm under the force of one volt. The relationship between current, voltage, and resistance in a closed circuit is given by Ohm's Law :

Current (amperes) = voltage (volts)/resistance (ohms)

The current in a circuit is directly proportional to the applied voltage and inversely proportional to the circuit resistance. That is;

$$I = V/R,$$

where,

I = Current in amperes (amount of electron flow),

V = Voltage in volts (amount of charge causing electron flow), and

R = Resistance in ohms (restriction of electron flow).

The flow of current in a circuit is like the flow of water in a pipe. The pressure (voltage) drives the flow (current) through the pipe (circuit). The amount of flow the pipe can handle depends on its size and material (resistance). As the flow reaches the end of the pipe, it releases energy to do work at some rate (power).

Only two of the three Ohm's Law quantities are needed to calculate power :

W = VI,

= V square/R, and

= I square R

Wattage is simply the product of voltage and amperage.

In electrofishing, the Ohm's Law parameters are redefined in three-dimensional terms. In electrofishing, a closed circuit is created by passing electric current between two submerged electrodes through the water and fish. Current of sufficient densities will either frighten, lead, stun or kill fish. As current flow leaves the electrodes, passing through the water, it spreads out in all directions forming a field pattern. Ohm's Law parameters for water now become voltage gradient, current density and resistivity.

Resistivity (ohms/cubic cm) = voltage gradient (volts/cm)/current density (amps/cm square).

Current density can be visualized as a measure of intensity of electron flow (current) at a given point in the water. The voltage gradient is the voltage between two closely spaced points causing the electron flow between the two points. Resistivity is the measure of the quality of the water as an electrical conductor. Resistivity is often referred to as conductivity and is the inverse of resistance (Smith, 1989).

Electrical Current Types

The current types used for electric fishing can be divided into two main types:

1. Bipolar or alternating current (AC), characterized by continually reversing polarity.
2. Unipolar or direct current (DC), characterized by movement of electrons in one direction only.

DC can be further sub-categorized into continuous DC (CDC) and pulsed direct current (PDC). For all types of current the pattern of voltage and current around the electrodes conforms to the pattern shown in the Figure 1.

When the fish are aligned along the current lines they will experience the greatest voltage potential, when aligned along the voltage lines they should experience the least voltage difference. However, they will experience some lateral voltage gradient across their body.

Alternating Current (AC)

The current direction reverses many times a second thus there is no any polarity to the current (one electrode being successively positive and negative many times a second. AC may be single phase or multi (usually 3) phases. This waveform has the advantage of being able to be produced easily from small generators and suffers little variation in effectiveness due to physical parameters of the stream (streambed conductivity, temperature *etc.*). The voltage gradient required to provoke a reaction is also quite small. When fish encounter an alternating current (AC) field they experience;

Oscillotaxis

The fish are attracted to the electrodes (but not to the same extent as with DC and PDC).

Transverse Oscillotaxis

The fish quickly take up a position across the current and parallel to the voltage lines in order to minimize the voltage potential along their body.

Tetanus

Once so aligned the fish muscle are in strong contraction and the fish are rigid. Breathing is also impaired by the fixation of the muscles controlling the mouth and opercular bones. The effect is more violent than with DC or PDC and at high voltages muscular contractions may be so severe that the vertebrae are damaged. The recovery time can be significant.

The disadvantages of AC are predominantly that it has minimal attraction effect and its effect upon fish is to tetanize the fish with its muscles in a cramped state. This tetanus quickly restricts the fish's ability to breathe and renders them unconscious. If not removed quickly from the field, death may occur quite soon from asphyxia. Delayed mortality may also occur due to acidosis resulting from the oxygen debt generated by the contracted muscles. Kolz (1989) found that even when applying the same power to the fish, fish immobilized with AC took longer time to recover than fish immobilized with PDC. In addition, with little attraction to the electrode, fish are not drawn out of cover or deep areas to where they can be seen and caught.

Direct Current (DC)

This is the simplest waveform used and technically is not a true "wave" but a constant voltage applied over time. The electrical charge flows only in one direction;

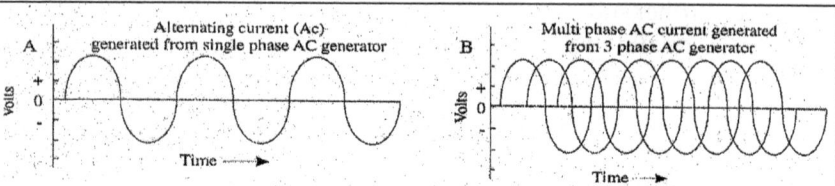

Single phase (A) and multi-phase (B) ac current pattern

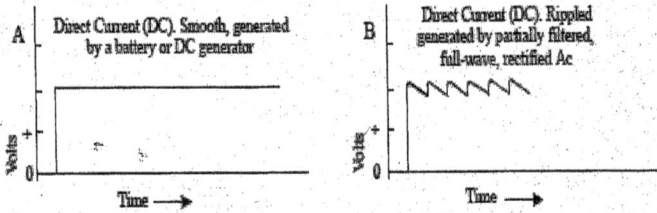

"True" (A) and "rippled" (B) direct current

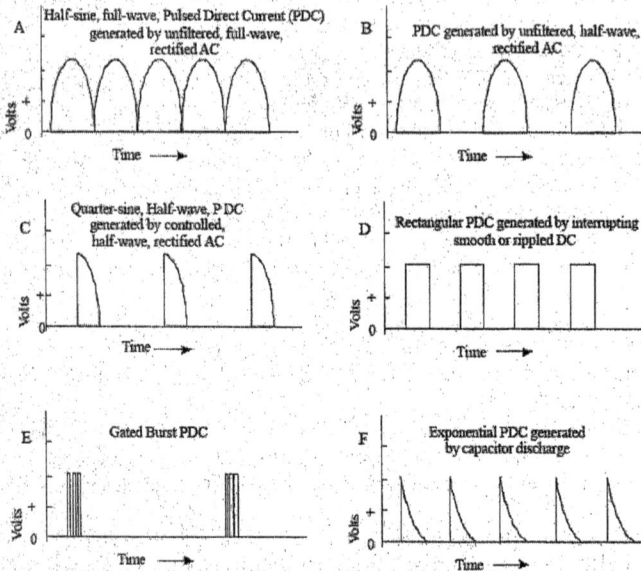

Examples of a range of pdc waveform types

Figure 1: Electrical Current Types.

from negative (cathode) to positive (anode). Direct current was the first type of electrical waveform to be applied to electric fishing; this is because it is the type that is produced from a galvanic cell (battery). Generating it needs a considerable amount of power however, thus requiring large generators or quickly exhausting batteries. Generators designed to produce DC current are heavier, more expensive, less reliable in voltage control and less reliable than AC generators with comparable power rating. For these reasons DC power is usually produced by conditioning power from an AC generator. In the past this conditioned DC often had a noticeable ripple resulting from inefficient smoothing of the AC source current; modern electronics however should give a good DC waveform. As the two electrodes (negative charge, cathode and positive charge, anode) produce differing physiological responses, the fish reaction will vary slightly depending upon which electrode it is facing. In field situations however the cathode field should ideally be very diffuse and thus should not influence the fish. Reactions to the anodic DC field can be broadly categorized into five basic phases.

Alignment

With initial electrical introduction the fish align themselves with the direction of electrical current. If initially transverse to the anode, the fish undergo anodic curvature that turns the head toward the anode.

Galvanotaxis

Once parallel with the current the fish start to swim towards the anode. This is achieved through electrical stimulation of the central nervous system resulting in voluntary swimming.

Galvanonarcosis

When the fish get close enough to the anode to experience a sufficient voltage gradient their ability to swim is impaired. In this state their muscles relaxed.

Pseudo-Forced Swimming

As the fish gets even closer to the anode, a zone where the fish begins again to swim toward the anode occurs. This swimming is caused by direct excitement of the fish muscles by the electric field and is not under the control of the central nervous system.

Tetanus

At high DC voltages the muscles go from a relaxed state into spasm. This can result in impaired ability to breathe and possible skeletal damage.

Unless held under conditions of tetanus, when the electricity is switched off, or the fish are removed from the electric field, they recover instantly. DC has a far greater attractive effect than other waveforms (AC and PDC) but it is less efficient as a stimulator and thus will not narcotize/tetanize the fish so readily. This is because threshold values required to elicit responses are high with DC compared to AC and PDC. As it also shows great variation in effectiveness for slight variations in the physical factors that affect it, any physical factors, which may affect the DC field characteristics, are likely to substantially reduce the effectiveness of the process.

Pulsed Direct Current (PDC)

This waveform is like a hybrid between DC and AC. It is unidirectional (*i.e.* it has no negative component) but it is not uniform. It has a lower power demand (like AC) but is less affected by physical variations in stream topography (unlike DC). Voltage gradients required to elicit a response are also substantially lower than those for DC. The shape and frequency of the pulses can take many forms, some of which are better than others with regard to their effectiveness and the injuries they cause. The behavior of fish to PDC is somewhere between that of DC and AC. As with DC the fish react differently to the anode and cathode field and thus their reaction will vary depending on which electrode they are facing. There are debate as to whether PDC produces true galvanotaxis and whether narcosis or tetanus causes immobilization. In general terms however a fish's reaction to a PDC field can be summarized as follows.

Electrotaxis

There is a good attracting power but this is due to the electrical effect on the fishes muscles (the muscles contracting with each pulse of electricity and thus accentuating the swimming motion) and not as in DC, by electrical effect on the spinal nerves. This vigorous effect upon the fish can also increase injury rates.

Tetanus/Narcosis

Like DC the fish are immobilized near the anode but at a much lower voltage gradient, as tetanus may be involved, the fish need to be removed from this zone quickly.

Pulse Frequency

Frequencies of pulses are measured in pulses per second or Hertz (Hz). Within UK only two pulse frequencies are commonly used (50 and 100 Hz). The principal reason for this is historic in that originally the source of the electricity was a commercial generator (producing 50 Hz AC) and the pulse box either full wave rectified the AC (producing 100 Hz PDC) or half wave rectified the AC (producing 50 Hz PDC). In the USA however the equipment used enables a wide variety of pulse frequencies to be used and considerable experimentation has taken place regarding the most efficient pulse rates to capture different species. Justus (1994) and Corcoran (1979) finding that optimal frequency even varied between similar catfish species. Novotny and Priegel (1974) stated that some species selectivity is possible by varying the pulse frequency of PDC. Halsband (in Vibert, 1967) stated that the frequencies mentioned in the table below were optimal for tetanizng those species. It may be noted however, that it may not be desirable to produce tetanus as it is the frequency to produce the greatest attraction effect that should be optimized (Hickley,1985, 1990).

Optimal Tetanizing Frequencies for different Fish Species (Halsband, 1967)

Species	Optimal Frequency (Hz)
Minnow	90
Trout	80
Carp	50
Eel	20

Electrophysiology

The basic principle of electric fishing is the transfer of electrical current into the water via electrodes and through the fish at enough current densities to produce a desired effect (taxis, repulsion or death). It is possible to stimulate or catch fish with any kind of electrical current (of a sufficiently strong field), but in order to maximize catch-per –unit-effort (CPUE), to avoid causing injury to fish, or to fish under adverse physical conditions, the proper choice of electrical parameters and current is important. There are three types of current; alternating current (AC), pulsed direct current (PDC) and constant direct current (DC).

It is well established that AC can efficiently tetanize (immobilize) fish. A serious side effect of AC is the potential to kill a high percentage of affected fish. Unlike DC, current direction is changing every half cycle. In an AC field, the fish faces the cathode and anode successively as many times as the current alternates (Lamarque, 1990). Above a certain field strength, this continuous reversing of current polarity quickly overwhelms the fish's nervous system. Constant DC has the desirable characteristic of producing anodic galvanotaxis (forced swimming toward the anode) with less harm to the fish. However, constant DC has a more limited effective range and generally large and efficient power requirements.

At the same peak power, AC, DC, and PDC will have similar or equivalent fields in terms of size and intensity. However, the response threshold levels of fish are higher for DC, thereby reducing the "effective zone". Also for DC, peak power = average power, whereas for PDC and AC average power, which determines the size of the generator, is much less (D. Synder, Colorado State Univ., pers. Comm.). The effects of pulsed DC are intermediate between that of AC and DC. Pulsed DC is most commonly used in boat-based electric fishing.

In choosing the appropriate electric parameters, one need to understand the behavior of the fish in electric fields (electrophysiology). Unfortunately, fish electrophysiology is generally not well documented nor understood. Galvanotaxis is believed by some to be a result of direct stimulation of the central and autonomic nervous systems that control the fishes voluntary and involuntary reactions.

Many authors have classified fish reactions in electric fields, attempting to fully explain their causative mechanisms. A general agreement of the results has proven difficult to achieve. The one matter that most scientists agree is that AC is more harmful than DC. Pulsed DC can produce desirable effects of both AC and DC while limiting the negative side effects (Lamarque, 1990).

Fish may exhibit four general responses to induced electric fields (*i.e.,* PDC at 60 Hz), avoidance, taxis, narcosis and tetany. These responses depend on the total duration and level of current density experienced. When electrofishing, it is necessary to establish an electric field of sufficient current density to achieve the desired response from fish. The field established is defined by three zones of increasing density; the perception zone, effective zone, and danger zone. If the perception zone is too large, fish are frightened and avoid capture. The desirable effect of taxis (forced swimming) occurs within the effective zone. If fish are not removed in a timely manner, narcosis (an induced relaxation of the body) occurs. Fish exposed to the danger zone will experience seizer or tetany. Tetany is the rigid immobilization of all musculature. Fish become tetanized by the increased levels of current densities. Tetany most often causes death by asphyxiation. Ideally an electrofishing unit produces the smallest perception zones, largest effective zone and no danger zone.

A fish's first reaction to AC is to take up a transverse position to the electric field lines; oscillotaxis (Koltz and Reynolds, 1989). The fish then repeatedly attempts to face the anode and cathode until threshold current is reached, causing the fish to be tetanized on the spot. Some authors also describe movements toward, as well as away from the electrodes (Lamarque, 1990). Little agreement on results was apparent in the literature review of electrophysiology.

In DC electrofishing, electric current flows continuously from the negative cathode to the positive anode. The actual mechanism for electron flow is electrolysis, that is the movement through water of ions that collect electrons at the cathode and release them at the anode (electron flow). The reaction of fish to DC is quite different than to AC. The first reaction observed in a DC field is a quivering of fish body muscles or fins; this occurs as the fish enters the perception zone. What happens after a fish enters the perception zone depends on a number of factors; the fish's orientation to the electric field (facing anode or cathode), species electrophysiological characteristics (resistance, fatigue), and current density. Assuming the fish does not flee (the perception zone), it then moves into the effective zone. As the fish moves through the effective zone it experiences increasing current densities, causing inhibited swimming followed by galvanotaxis. If the fish is not removed from the increasing field densities, it will continue its forced swimming toward the anode until relaxation of all its muscles is induced (galvanonarcosis). With prolonged exposure to DC, a second forced swimming occurs, which sends the fish into the area of highest current densities, the danger zone. Here tetany occurs, often followed by death (Lamarque, 1990). If a fish is removed from the danger zone in time and allowed to recover under optimal conditions, death from tetany may be averted (D. Synder, Colorado State Univ. Ft. Collins, pers. Comm.).

Strong anodic taxis is possible with pulsed direct current. Lamarque (1990) suggests that the mechanisms of taxis are quite different for DC and PDC. PDC is produced by interrupting a steady DC current flow with an electronically controlled switch. The switch gives the number of ON-OFF pulses per second (frequency). Research has shown a species-specific reaction to frequency and pulse width. In general, pulse shapes with a fast rise and slow decay enhance anodic taxis. With

PDC, no narcosis or second swimming towards the anode occurs. In the effective zone, fish are drawn more directly from a greater distance than DC toward the anode, generally becoming immobilized before reaching the danger zone.

The establishment of the perception, effective, and danger zones in AC, DC, and PDC depends on field strength, water conductivity, and electrode size.

Summarizing, it can be said that DC produces galvanotaxis, inducing tetanus only near the electrode and after prolonged exposure, and is the least harmful to fish. However, it has the most limited effective range and highest power requirements. PDC produces strong galvanotaxis, and has a large effective zone and greatly reduced power requirements. PDC does tend to immobilize a large portion of the catch farther away from the anode than DC. AC has the greatest effective range but little or no taxis, with the potential for tetanizing fish, resulting in the death and loss of capture.

Circuit Principles

An understanding of the basic concepts of matter and its relationship to charged particles is the beginning. Matter consists of electrically charged particles, called electrons (-), protons (+) and neutrons. Although all matter contains electrically charged particles, most matter is at a neutral charge. Electricity is caused separating charged particles, and electrical energy is then created by the attractive forces between protons and electrons.

The second in importance is the transfer of electrical energy, which requires the creation of a circuit. A circuit is a closed, insulated system that allows electrical energy to travel along a particular pathway. In electrofishing systems, usually only the voltage is adjustable, but adjusting voltage often results in changes in current.

There are two kinds of circuits, series and parallel. In a series circuit, resistors, or loads are arranged sequentially in the circuit; if one load is removed, the whole system shuts down. In effect, the loads become switches, if they stop working. In this type of system current is constant through all the loads and voltage is variable assuming that the loads are different in size. Total voltage is equal to the sum of voltages at all resistors. The total resistance of an electrofishing circuit or "equivalent resistance" is the sum of all the individual resistors or loads. In a series circuit, this is the sum of the individual resistances.

Current is constant in the system but voltage varies if the resistors have different values. Changing the resistance in a system alters the relationship between voltage and current, current is generally reduced in a system with high resistance. The energy dissipates as heat in the circuit, overloading the wiring and the load.

Parallel circuits differ from series in that they have branched pathways. Each branch usually has a resistor, and the resistors are not sequential system as in a series circuit. The current splits at each branch. In the series circuit, current is constant and voltage varies. Parallel circuits are the opposite, voltage is constant and the current is variable at each load, when the loads have different resistance. When all the loads have equal resistances, the voltage is the same at each resistor. Overloading of one of the loads does not interrupt the total flow of energy, other loads continue to function. Parallel circuits have another advantage that constant voltage can be maintained, delivering 110 to 120 volts to each pathway. The current varies depending on the load. The equivalent resistance of a parallel circuit is calculated as the inverse of the sum of the inverses for all the system's resistors. Earlier (1950) scientists believed in the conventional flow theory, in which electricity flows from the positive to the negative and protons carry the electrical charge. Since then, physicists have proved that electrons carry the electrical charge and that the direction of flow is from negative to positive, a principle known as electron flow theory has been established.

Ohm's Law states that resistance is a constant ratio between the voltage and the current. The ratio between the voltage and the current remains constant, regardless of the voltage.

With this basic knowledge of circuits, it is possible to construct an electrofishing system at least on paper. The system has two anodes (positive electrodes) and one or more cathodes. For metal boats, the boat hull often serves as the cathode or dropper electrodes is used. The anodes are positioned of a boat's bow, or, for backpack shocking units at the end of the rod. The fishing and netting occurs at the anode.

If an electrofishing system were configured as a single series circuit, removal of any of the loads would interrupt the circuit, which implies that taking any of the anodes out of the water should cause electrofishing to stop. This does not occur because the anodes are in parallel circuit within a large series circuit. If there is only one cathode, however, removing it from the water interrupts the circuit and terminates electrofishing. If there are multiple cathodes, configured as droppers, for example, they occur in a parallel circuit in the same manner as the anodes, so that electrofishing continues when some droppers are not suspended in the water. This brief description applies to most electrofishing systems.

The basic concept for circuit lay out of an electrofishing system is straight forward It involves both parallel and series circuits. The overall resistance of the system can be calculated by first simplifying the parallel circuits into equivalent resistances which reduces the system to a series circuit in which the total resistance is the sum of the individual resistors. The basic difference between electrofishing systems and the circuits described here is that water replaces some of the wiring and the load of interest is the fish.

Example

An electrofishing system has two anodes with resistance of 75 and 50 ohms. The cathode is bigger and it has a resistance (Rc) of 10 ohms. Applied voltage is 200 volts. The equivalent resistance (Req) of this electrofishing system is determined by first calculating the equivalent resistance of the anode parallel sub-circuit (Ra), which is the inverse of the sum of the inverses; $Ra = 1/(1/75 + 1/50) = 30$ ohms. Then the total equivalent resistance is $Req = Ra + Rc = 10 + 30 = 40$ ohms, and the current $(I) = V/Req = 200/40 = 5$ amps.

The objective in electrofishing is to transfer energy from one load to another, from water to fish. The resistance of the fish is generally different than that of the water.The difference between the resistance of the water and the resistance of the fish can act as a barrier to energy transfer. The input effort in electrofishing is limited by the ability to control the available energy in the water. Factors regulating the energy transfer from water to fish are beyond human control, but understanding the factors allow to respond intelligently.

Another important concept in electrofishing is power (P), which is energy per unit time. Power is voltage multiplied by current, where voltage is energy per unit charge and current is charges per unit time. Cancelling out the charge leaves energy per unit time, Joule's Law is $P = I \times R$, or $P = V$ square$/R$.

Example

In the electrofishing boat for example, there was an applied voltage of 200 with 75 ohms resistance for one anode, 50 ohms for the other and 10 ohms for the cathode. The equivalent resistance for the system was 40 and the current for the system was 5 amps. The power is, $P = 200 \times 5$ amps $= 1000$ watts. Given that 746 watts equals one horse power (hp), 1000 watts equals 1.3 hp.

Waveforms

A trace graph of voltage in a circuit with an operable switch would show an instantaneous increase (because electrons travel at the speed of light), when the switch was closed and instantaneous decrease when the switch is opened. This is called direct current (DC). Electrons flow in one direction from the cathode to the anode, because the polarity never changes. The simple DC waveform is characterized only by the voltage. A trace graph of the current would be identical to that for the voltage.

For many years, DC was considered a good waveform for electrofishing as it was less harmful to fish than some other waveforms. Producing DC, however, requires maintenance of high power inputs with a large heavy generator. More recently pulsed DC has been employed in electrofishing. Pulsed DC is produced by the regulated interruption of the continuous flow of DC, which yields a pulsed signal. Pulsed DC is also effective at delivering energy to the water, and is less traumatic to fish than other waveforms, but it also has the advantage of not requiring continuous high-energy inputs. Most electrofishing systems do not produce a classical square pulsed DC waveform, because pulsed DC can be produced in a variety of ways. More circuit and system components add to production costs, but some pulsed DC waveforms are less expensive to produce than others.

Pulsed DC is more complex than DC. Waveforms are characterized by the maximum voltage (Vmax) and an average voltage. Current flows the same general pattern as voltage. Determining the average voltage is difficult, it depends not only on the maximum voltage but also on the duration and spacing between pulses, which is usually in milliseconds (m sec). In electrofishing 5 millisecond is a typical pulse time, referred to as the pulse duration or pulse width. The number of pulses per unit time is called pulse frequency, pulses per second or hertz (Hz). The typical frequency of an electrofishing system unit is 50-60 Hz.

It remains uncertain which factors are most important in fish response to pulsed DC. It may be average voltage or peak voltage, overall power, or a combination of all of these variables. It appears that peak power and peak voltage are important. Fish do seem to respond to peaks, but if the peaks are not wide enough, they do not respond at all. Fish response, therefore, must be a function of some combination of peak power and average power, where power delivered is determined by both voltage and current. Another type of waveform is alternating current (AC) which is characterized by a switch of polarity. AC is the waveform output of most generators. The direction of the polarity changes at the generator's speed, which is usually 60 Hz (60 cycles AC). If AC is used in an electrofishing system where the two electrodes are the same size and shape, both electrodes have the same fishing effect. Fish are not drawn by one electrode more than the other.

AC waveforms are characterized by peak voltage and peak to peak voltage which is the difference between the positive peak and negative peak. The average voltage is meaningless, because adding the positive side and the negative side yields an average of zero. Therefore, root mean square (RMS) voltage is used to characterize AC waveform. RMS voltage (rms) is the most common output format on electrofishing and voltage meters. A typical sigmoid AC wave form produced by a generator has a peak voltage (Vp) equal to RMS voltage (Vrms) divided by 0.707. The peak to peak voltage is 2xVp or 2x(Vrms/0.707) =2.6. Therefore peak to peak voltage is nearly three times the Vrms voltage. Vrms voltage are common on electrofishing units, which is very detrimental to fish because of the changing polarity. A fish orients to the positive in a DC system. In an AC system, fish can not orient to the positive 60 times a second which is the rate of polarity switches in a typical electrofishing system. The fish enters oscillotaxis, characterized by quivering and unpredictable reactions. Sometimes, in a low conductivity waters, the only way to catch fish with electrofishing is to use AC. To achieve the same results with pulsed DC would require a bigger generator to put more energy into the water, which sometimes is not feasible. The general recommendations in electrofishing, however, is to avoid using AC if possible, because it poses a greater threat of injury to humans and fish alike.

A relatively inexpensive method of producing pulsed DC is to rectify the AC waveform. By cropping the waveform to only the positive half, a pseudo-pulsed DC waveform is produced at the generator's cycle speed. A 60-cycle generator would produced 60 Hz pulsed DC with rectified AC form. To fully rectify an AC form, the negative portion of the waveform is flipped to the positive. This produces a pulsed DC form because it returns to zero after each pulse, and is always positive. It also double the frequency.

Most electrofishing units take AC from the generator and produce a continuous DC, then break it up into various patterns. This function provides the most flexibility in terms of frequency, duty cycle, pulse duration and amplitude. The biggest limitation of these systems is the inability to vary the pulse shape. The choice of pulse shape influences cost of production; exponential decay waveforms are fairly inexpensive to produce. Among pulse shapes, a square wave has the greatest area under the curve, and thus maximized power output for a given voltage. Other wave forms have lower average power or average voltage given the same peak.

Minimizing the area under the curve, or using less power, is one way to minimize injury to fish. One of the most popular approaches at present to emit "pulse trains" or a "pulse energy packet" or "a low-energy waveform" rather than rectangular pulses. These newer waveforms are characterized by short duration spikes, are clumped in time, and are produced at a high frequency (greater than 200 Hz). The "packets"are then produced at low frequency (10-20 Hz). The result is a low frequency pulsed DC with less energy in each packet than in an equivalent pulse. The objective is to reduce the rate of severity of injury by eliciting the same response from the fish using less energy.

Most electrofishing units do not provide a means of viewing emitted waveform, this is useful to examine the output waveform with an oscilloscope. The equipment should be checked with an oscilloscope to permit detection of AC effects, such as, a negative spike in a pulsed DC wave form. Not all pulsed DC electrofishing systems use a true pulsed signal, and changes in equipment and components that occur with aging can alter emitted waveforms. Battery powered units are more prone to AC component waveforms. The On-Off switches of the pulse causes an overshoot. Correcting this problem is expensive, and the problem is not usually present in older backpack units. Only an oscilloscope will reveal whether (or not) a pulsed DC waveform has negative spikes. On most scopes, overshoots appear as a very faint line.Overshoots can be 2-3 times greater than the pulse in magnitude, but are usually of short duration (microseconds). Fish can be harmed by these overshoots (negative spikes). Using traditional pulsed DC systems (not low energy waveforms) to catch adult fish (> 20 cm), the duty cycle should be maintained at 50 per cent or lower. A duty cycle of 10-15 per cent would be preferable if it were successful in catching fish. Ideal frequencies are 20-25 Hz, and frequencies above 100 Hz should be avoided because they will harm adult fish of all species. Frequencies of 50-60 Hz are very injurious to adult salmonids. Most backpack and boat units provide a 50 per cent duty cycle and 50 Hz as a "standard" waveform; it is desirable to be able to control not only voltage, but also frequency and duty cycle (or pulse duration).

Fish Reaction and Sensitivity

All fish have approximately the same nervous system, which are similar to those of other vertebrate species. In a dorsal view, the nerves follow the myomeres out from the spinal cord and integrate the muscles. The front of the brain appears to carry a negative charge, which may explain why fish are attracted to anodes. The nerve angles and lengths appear to be important determinants of electrofishing, as the orientation of the fish to the electrical fields determines how the fish will be affected.

The objective in electrofishing is to interfere with this neurological pathway between the brain and muscles of the fish. By blocking the internal signal and overridding it with a signal from the water, electricfishing currents redirects the neurological signal and muscular reaction. Fish can swim undisturbed through an electrical barrier if they are oriented in a particular position in the electrical field.

Likewise, once in the electrical field, the behavior of a fish varies with its orientation. The optimal reaction is involuntary swimming in a predictable direction (towards the anode). If the force of muscular contraction is too great, however, spinal injuries may occur.

Electrofishing is size selective. Larger fish tend to be more vulnerable as their length spans a greater voltage gradient (referred to as head to tail voltage). Considering the pulse traveling in a wave length; a larger fish intercept more wavelengths and more energy. There is also a dorsal-ventral vector. It is important to think in terms of power, rather than voltage or current, as the power encompasses the waveform in the water. The power waves goes through the water too, and affects the fish from different vectors. Although the mechanisms are not completely understood, it is known orientation of fish in the field is also important.

There is an important distinction between size selectivity in capture efficiency and mortality. While capture efficiencies usually increase with fish length, mortality is more closely related to responses to pulse frequency and duration. Differences in skeletal and muscular structure among species and size classes may influence susceptibility to injury. Another important difference between species is the extent and nature of scales. For example, salmonids are very finely scaled and have little resistance to energy transfer. A heavily scaled fish, such as, carp, on the other hand, is more resistant at the same energy level. Proportion of muscle mass relative to the total body weight may also be a factor. Weak swimmers have a very low proportion of muscle mass, probably below 50 per cent. Strong swimmers, such as, migratory species, have well above 50 per cent muscle mass. In salmonids, the proportion is high (60 per cent). A higher proportion of muscle mass increases the response to applied electricity.

Many unanswered question remains. The stability of the conductivity of the fish is not known, but is assumed to be constant for a given species and size group. According to power theory, energy transfer is greatest when the conductivity of the fish is similar to that of the water. While the conductivity of water can be readily measured, conductivity of live fish can only be estimated.

It is not fully understood the relationship between power transfer theory and pulsing waveforms. In electrofishing, however, it is useful to think of energy in terms of power, which is energy per unit time, in much the same way as we refer to energy for home use as watt hours, also energy per unit time. Power, not pulse length, dictates fish response. In general terms, using energy, but it is actually power that pertains to electrofishing effects.

Fundamental Electrofishing System

The basic lay out of an electrofishing system is given below. In battery operated units, the battery produces continuous 12 volt DC power. If the electrofishing unit served only to interrupt the power, a peak voltage of 12 volts would be produced. But this is not enough power to stun fish. The power either from a battery as pulse DC or from an AC producing generator, is then put through a transformer, which enhances the input signal to produce a higher output signal. Pulsing the battery signal is necessary to prevent the transformer from burning out. The transformer treats pulsed DC signals in the same manner it treats AC signals. The transformer produces a high voltage AC signal. For AC electrofishing, this is all that is required. Therefore, the simplest systems are composed of a battery, a DC : AC inverter and a small transformer. These systems are not very powerful, but they do create an AC signal suitable for electrofishing. Routing the electricity through an AC to DC converter provides a high DC signal. The addition of a pulser provides a high pulsed DC waveform.

Battery ——— **DC-AC inverter** ————————————— **Transformer**

Generator

High voltage   **High voltage**

                                            Could fish

Power to water ——— **Pulser** ——— ———— **AC – DC converter**

Could fish

The Basic Layout of an Electrofishing System

Maximum Power Transfer

In circuits, there are two types of loads or resistances, the internal resistance of the system and a load resistance. A critical concept in electrofishing is maximum power transfer, which occurs when the internal resistance is equal to the load resistance.

Example - A series circuit has an internal resistance (Rg) of 10 ohms and a variable load resistance (Rv). The applied voltage is 120. The following equations for circuit apply;

P (power) is equal to I square x Req,

Req = Rg + Rv

I (current) is equal to V/(10 + Rr)

Vv = I x Rv

Pv = I x Vv

The following table demonstrates that in a system with a constant internal load, the power produced is variable in a non-linear pattern. As resistance increases, current decreases and voltage increases. Power, however, increases and then decreases, illustrating the principle of maximum power transfer. The maximum power transfer occurs when Rv is equal to 10. Maximum power transfer occurs when the load resistance is equal to internal resistance which is a circuit phenomenon. This principle applies directly to electrofishing, the variable load is the fish and the load is the water. In real situation, however, both loads are variable.

Rr (Resistance)	Maximum Power Transfer Determination		
	I (Current)	Pr (Power)	Vr (Voltage)
0	12	0	0
5	8	320	40
10	6	360	60
14	5	350	70
20	4	320	80
50	2	200	100

For practical purpose, the resistance of the electrical system, whether it is a backpack or a boat, is nominal relative to the water. The significant resistance is in the water and in the transfer to the fish. When the "resistances" (*i.e.* conductivities) of the fish and the water are equal, any power put into the water will be fully transferred to the fish. In electrofishing practice, it is not possible to know if a matched condition exists. In some instances, even if conductivities are equal, the power delivered to the fish may not be sufficient to elicit the desired reaction. This principle, however does not explain why fish can be caught on one occasion, but not in another in the same place and with the same equipment. The matching of the conditions may have changed enough that the power put into the water has effectively dropped below transfer threshold from water to fish. This usually occurs when the water conductivity changes. An electrofishing system may work well at 360 watts on one occasion, but then on a repeat visit to the site more wattage may be needed. This phenomenon could be caused by a change in water conductivity related to rainfall, water temperature or time of day, which changes the ratio of fish and water conductivities.

The idea of achieving constant power is important for establishing electrofishing guidelines. Assuming that fish conductivity is constant, knowledge of water conductivity and temperature are required to standardize the power level. The power in the field can be determined using the following graph. The Y-axis is the per cent of maximum power transfer. The X-axis is the ratio of resistance load to internal load (*i.e.* ratio of fish and water conductivities). At a ratio of 1, power transfer is 100 per cent. At ratios of 2 or 0.5, 90 per cent power transfer is obtained. Although it seems that power transfer between low conductivity water and high conductivity fish, ought to be maximum, this is not true. Fishing at high conductivity is as hard as fishing at low conductivity, assuming the range of water conductivity is much wider than the range of fish conductivity. For practical purposes, fish conductivity is treated as a constant of about 150 mohms/cm based on the previous research of all fish species. Although this constant was determined in gold fish, and may not be representative of all fish species, it has been helpful in evaluating the importance of standardizing for power. By standardizing power to water conductivity, as much as 15 per cent of the variability in catch per unit effort (CPUE) can be explained. Given the variability in CPUE can vary considerable among samples, it is important to reduce variable to the extent possible through standardization of equipment and procedures.

People electrofishing in low conductivity water (< 50 µS/cm) tend to be "volt" people, because the current meter does not respond in low conductivity water. People who fish in high conductivity water (>150 µS/cm) tend to be "amp" people, because the volt meter does not change much. At low conductivities, there is very little change in current density, but a large change in voltage gradient. At high conductivity, there is little change in voltage gradient lines, but a large change in current density. Regardless, one fish with power – the product of volts and amps – which is why understanding power transfer theory is so important to understanding problems in electrofishing.

| 0 | 0.1 | 0.2 | 0.3 | 1 | 2 | 3 | 5 |

Ratio of Conductivities – Fish/Water

The amount of power that must be added to match a previous electrofishing effort can be determined using the graph.Inverting the Y-axis changes the percentage to a ratio, which is termed the power correction factor (PCF). The PCF is the additional power needed over the present situation to achieve a given power transfer. The equation for PCF is;

PCF = $(1+q)2/4 \times q$, where q is the ratio of fish conductivity and water conductivity.

To illustrate, one has been fishing in perfectly matched condition at 320 watts and returned to find a mismatch. In other words, one had been fishing successfully and assumed a match of 1, but upon return, the water conductivity was two fold higher. The additional amount of power required to achieve matched conditions would be calculated as the wattage (320) x the current matched power ratio (1.1). Therefore, 352 watts would be required to achieve the same power transfer. While standardizing the power transfer among sampling efforts is very important, actually doing so is often beyond the capabilities of existing electrofishing equipment. It may not be possible to adjust the voltage as precisely as would be required in this example. Nonethless, testing the settings prior to starting a depletion estimate study is extremely important to reduce variability. Also population estimation models typically do not allow for changes in settings among passes.

Checking the Electrofishing System using an Oscilloscope

Internal system aging manifests itself in an electrofisher as the need to change settings to achieve the same effect. When a system drifts beyond the adjustment capabilities of the unit, internal parts may need to be replaced.

An oscilloscope should be used on electrofishing units once a year to calibrate settings and output meters to the waveforms emitted, and to detect defects. It is critical that the oscilloscope be supplied with independent power source, and not run off the same generator that is powering electrofishing unit.

The output of an oscilloscope can be used to calculate frequency, as well as peak and average voltage. The goal of electrofishing is to use minimal amount of energy required to elicit reaction from the fish. It is best to start with a low frequency and a fairly narrow pulse width and increase the amplitude as necessary until the fish react. If this does not work, start over with a higher frequency until results are obtained. It is a good idea to compose a table of setting voltages vs real voltage and attach it to the unit. It would even be possible to connect an oscilloscope to the anode and cathode during electrofishing to monitor the power waveform. Power density determines the effective range of electrofishing systems. It can be determined using a field strength or voltage gradient meter. While many fish biologists feel that such procedures are excessively time-consuming, undertaking these analysis prior to field season can significantly improve the interpretability of electrofishing data.

Although it is not possible to know directly if the maximum power transfer condition is occurring during electrofishing, it is helpful to monitor the fish reaction at different voltage settings and determine the power output (voltage x current). From this it is possible to build up a database for water and species in an area of interest.

One source of variability is the equipment. Electrofishing units of the same model from the same manufacturer may emit significantly different waveforms (even if the products meet the manufacturer's standards), probably as a result of change in components during the production period. Therefore, every unit purchased should be characterized with an oscilloscope.

A Typical Cathode Ray Oscilloscope

The cathode ray oscilloscope (CRO) is a device which converts electrical energy into a graphic image, called the trace. The heart of the device is a glass tube with a screen like a TV. When an alternating current (AC) is applied to the CRO inputs, it shows up as a simple waveform on the screen; direct current (DC) is displayed as a straight line. The CRO is a good device for measuring voltage because it has a very high internal resistance (> 1 megaohms) and does not interfere very much with the operation of the circuit under investigation. The trace can be analyzed in terms of voltage (vertical axis measurements) and time (horizontal measurements).

Setting up the Oscilloscope

Shortcut method to obtain a trace on the scope.

1. Turn the scope off.
2. Turn all knobs with a CAL or AUTO setting to CAL or AUTO.
3. Turn all other knobs to the mid position.
4. Turn the brightness knob to maximum.
5. Switch the scope on and wait for about 2 seconds to 1 minute as most scopes need time to warm up.
6. If the trace does not appear as a straight green line or a moving spot, follow the detailed setup methods below.

Detailed Method to Obtain a Trace

1. Identify the POWER switch, often coupled with the brightness control and turn it off.

2. Turn the FOCUS control to the mid-way position.

3. Rotate the TIME-BASE selector switch (coarse control) to a suitable setting. This adjusts the time taken for the spot to move 1 cm across the screen. The frequency of the waveform will be reciprocal of the time taken for one complete cycle. For a 50 Hz signal from a power pack choose the 5 ms/ cm setting on the TIME-BASE selector so that 1 cycle will take 20 ms and cover 4 cm of the screen. Only select EXT or HOR INPUT when feeding a signal to the horizontal axis as well as the vertical axis.

4. Rotate the fine control TIME-BASE knob to CAL (calibrated) which is necessary if accurate voltage readings are to be taken from the screen.

5. Set the DC/AC switch to DC and leave it there (AC is only used when trying to view the small varying component of a DC voltage.

6. Turn the X-SHIFT and Y-SHIFT knobs to the mid-position.

7. Switch the X-GAIN to the CAL (calibrated) position.

8. Adjust the Y-GAIN to a mid-way setting such as 1 V/cm.

9. Set the TRIG level to the auto position.

10. Set the TRIGGER +/- switch to the + position and set the TRIGGER source to the input selected for use; CH 1 or CH 2.

11. Turn ON the power and allow for a 2 minute warm-period.

12. Rotate the BRIGHTNESS control to a maximum.

13. Adjust the STABILITY control so that the trace just starts to appear on the screen.

Further Setting up Procedure

1. Connect the voltage to be measured by the CRO to the Y-INPUT (VERT INPUT) consisting of two sockets, the lower one of which is earthed (-).

2. When connecting a DC power supply to the Y-INPUT of the CRO, always connect the negative terminal of the power source to the EARTH input socket of the CRO.

Practical Experiences

1. To investigate the nature of AC and DC voltages – For DC, connect a 1.5 V dry cell to the Y-INPUT terminals of the CRO. Adjust the Y-GAIN to 0.2 V/cm and the TIME-BASE selector to 1s/cm. For AC connect the power source to the Y-INPUT terminals of the CRO. Set the Y-GAIN on 5 V/ cm and the TIME-BASE selector on 100 ms/cm. Note the waveforms and amplitude.

2. For investigating both the AC and DC voltage output from a power supply, connect the AC terminals of a power pack unit to the Y-INPUT terminals of the CRO and adjust the setting to "B". Set the Y-GAIN on 2 V/cm and the TIME-BASE selector on 10 ms/cm. Switch to the DC terminals of the power pack and observe that the so-called DC current is really rectified

AC- it is not a smooth consistent voltage output but contains a ripple effect.

Electrical Field Theory

Most electrofishing studies have focused on electrofishing effectiveness in terms of circuit theory. Only recently there has been an emphasis on the characteristics of the electrical field itself. In a three dimensional medium, distance becomes important. The resistance is directly proportional to the length and inversely proportional to the inside surface area (cross section). The resistance indicate that measurements occur in a field rather than in a circuit. Resistance is measured in ohms. Resistivity (electrical friction) units are ohm/cm. Just as the inverse of resistance is conductance, the inverse of resistivity is conductivity. Conductivity is a measure of how well electricity flows through a three dimensional medium, and is usually measured in Seimens/cm (S/cm). Because the ion concentration in water is low, conductivity is typically measured as micro-Seimens per cm (μS/cm). The two factors that affect water conductivity are ionic strength and temperature. As ionic strength increase, so does the conductivity. Conductivity affects current density, but has no effect on voltage gradient, if applied voltage remains constant. The ambient conductivity is the important parameter in electrofishing, but it is also essential to record water temperature, because fish physiology is sensitive to temperature.

Other important parameters in field theory are voltage gradient or volts per cm and current density, or current expressed per square cm. Current density is the flow of charge carriers through a plane. This is the second form or field form of Ohm's Law. Electrical fields in water have a very strong non-random orientation.

Power density is another important concept in electrical field theory. The product of voltage gradient (e) and current density (i) is equal to power density (D) or watts per cubic cm. Power density is the power dissipated in a three dimensional medium (D=exi). Neither power density nor current density can be measured directly, but they can be calculated from field measurements of voltage gradient and conductivity. Current density is equal to (voltage gradient)^2 x water conductivity. With an electrical field in water, water conductivity is typically constant because it is a function of water temperature and the amount of suspended solids. Water conductivity, however, will vary as the chemical composition of the water varies. Tributaries or point-source discharges can affect the conductivity of receiving waters. By contrast, the voltage gradient differs with distance from the anode. Maximum voltage gradient is used to derive current and power densities.

In most systems, half the total voltage is achieved at the mid-point between the anode and the cathode. The voltage gradient, however, is no linear, but changes more rapidly near the electrodes. A plot of the voltage between two electrodes would be symmetrical because the two electrodes are the same shape and size. Energy dissipates from the two electrodes at the same rate in terms of distance. Although the voltage gradient is the same at the cathode and anode, this does not mean that fish react in the same manner, the neuro-physiology of the fish directs them toward the anode.

The rate of change in voltage over distance is the voltage gradient. Voltage gradient measurements in water are symmetrical with the highest values at each electrode, and lowest value half way between electrodes. Depending on the spacing of the electrodes and the strength of the field, the voltage gradient can be zero in some areas. Mapping of voltage gradient at different distances from the anodes is an important performance evaluation technique in electrofishing. If anodes and cathodes are separated by too great distance, then areas of inadequate power density result. Also a rough rule of thumb is that effective electrofishing occurs at voltage gradients of 0.1 to 1 V/cm in water of moderate conductivity (200-300 µS/cm).

Voltage Data taken between Two Electrodes 25 cm Apart

Distance from First Anode (cm)	Volts (V/cm)
0	0
1.3	6.5
2.5	7.8
5	9.3
7.6	10.1
10.2	10.8
12.7	11.3
15.2	11.8
17.8	12.5
20.3	13.2
22.9	14.1
24.1	16.6
25.4	23.8

Below 0.1 V/cm, fish tend to escape and above 1.0 V/cm, fish tend to traumatized. For any given applied voltage and electrofishing configuration, field strength or voltage gradient mapping remains constant, despite changes in water conductivity. Voltage gradient at any point in the electrical field is directly proportional to applied voltage.

Although the mapped voltage remains constant, the power density changes as a function of conductivity. The characteristics of the field change whenever the position or orientation of the electrodes is changed, which may occur, for example, when cathodes drag along the stream bottom. This can occur with changes in bottom substrate and water velocity.

Electrode Configuration

In a pulsed DC system, the practical depth limit of a dropper acting as anode to attract fish is about 0,5 meters. The goal is to bring fish to the surface so they are easier to see and net. Droppers are more effective at greater depths for AC systems, and also when the anodes in a pulsed DC system are not effectively attracting fish. With an AC system in a very low conductivity water, steel cable or cylindrical droppers on the stream or lake bottom in 10-15 feet of water can stun fish, causing them to float on the surface.

An effective application for deep cathodes is in deep sided reservoirs, where a 10-15 feet cable, the last foot of which is not insulated is dragged along the bottom. This agitates fish from deeper water into the anodic field, where they are then attracted to the anode, which is closed to the surface.

A general rule is that cathodes should have a surface area equal to or larger than that of the anodes to prevent wasting power to the cathode in a DC system. The cathode ideally should be much larger than the anode. In backpack shocking units the small "rat tail" cathodes may have a smaller surface area than the ring anode, in this case, the cathodes are not optimal because they dissipate too much power at the cathode. One way to improve the performance is to increase the size of the cathodes. Screwing two cake pans together with the Styrofoam in the middle is an inexpensive and practical technique for increasing cathode size for backpack units.

Electrode design is a very important but often overlooked feature in electrofishing systems. Different electrode designs fulfill different objectives. A general guiding rule is that electrodes should be made of material as large in diameter as possible, given the voltage inputs. An electrode made from larger diameter material (cable or chimes,hollow cylinders) provides a bigger field that is less damaging to fish. Anodes are usually configurated as cable or chime dropper 3 mm in diameter produces a very intense field surface area of material. A single dropper 3 mm in diameter produces a very intense field close in, with almost no field further out, resulting in large dead areas. The result is that a fish does not react until it gets right next to the anode, then it feels the full force. Therefore, single droppers are not recommended except when fishing is extremely high – or low –conductivity waters, where minimal anode surface area is required to prevent generator overload. If generator overload is not a concern, then droppers of 12-25 mm are advisable. Increasing the diameter of the droppers decreases the intensity of the field near the anodes and extends the distance of the effective field.

Anodes can be of variety of shapes and sizes. A sphere provides uniform (multi-directional) energy dispersal. Large diameter spheres, however, are impractical in most situations because of the weight and drag they impose. An alternative is a large ring with droppers, often referred to as Wisconsin ring. In the 1960s, Novotny and Priegel (1974) conceived that the addition of anodes suspended from a ring would result in an electrical field similar to that generated by a sphere, because the weak individual fields generated by the droppers would overlap. The disadvantage of Wisconsin rings is that the sphere effect does not occur close to the droppers, therefore, the problem of intense fields close to individual anodes still applies. If fish are netted as soon as they enter the effective field, then this problem is not encountered. However, sweeping a net of fish close to the anodes can expose the fish to the full effect of an intense field.

The effective depth of anodes is determined by the fact that charge carriers are emitted only from metal surfaces perpendicular to that surface. For cylindrical droppers, the electrical field is only as deep as electrodes, because the surface area on the bottom of the dropper is very small.

Figure 2: Wisconsin Rings as Catcher Electrodes Operating from an Electrofishing Boat in a River.

The distance between the anode and the cathode is usually not important. When the anode is effective at attracting fish, the cathode acts only as return circuit for the system. For bank side shocker systems in which the generator is on the bank, the cathode can be buried to provide a good earth contact. When the anode is ineffective at stunning fish, decreasing the distance between the anode and cathode can increase field strength by overlapping the two fields. There is a danger in doing so, however, because equipment damage is likely to occur if the anodes and cathodes touch.

Safety

There is an effort to implement a nationwide inspection and certification program for electrofishers in some countries for safety and standardization process.

More fish biologists have been killed driving to the work site than have been electrocuted. More fish biologists also have been injured by the scare of an electric shock, which caused them to fall and hurt themselves in the boat on the ground than from the shock itself. In the USA, since World War II, only about five electrocutions during electrofishing have been documented. Nevertheless, electrofishing can be hazardous, therefore, safety must always come first. In other parts of the world-China for example, where systems are extremely primitive and people are untrained, the mortality rate is much higher.

The most hazardous type of operation in electrofishing is bank side shocking. Most bank side shocking systems include multiple anodes. The operation is usually large, and untrained personnel may volunteer to operate one or more of the anodes. That is how most people have been killed while electrofishing.

Backpack shocking and boat shocking operate on different safety considerations. In boat shocking, an equipotential surface should be maintained inside the boat. In this condition, a person can touch two metal objects in the boat and not be affected, because everything within the boat has the same potential. In backpack shocking, it is not possible to maintain equal potentials, therefore, insulation is the only form of protection. The leader of an electrofishing effort should be able to affirmatively answer the question, " If an accident occur, could I defend my procedures ?" in front of an objective investigative body.

Many electrofishing crews use anodes as dip nets. While this practice can be effective and successful if fish are immediately transferred to an assistant netter with a non-conductive handled net and insulated gloves, failure to remove the fish immediately from intense electrical field can be harmful to the fish, because the collected fish stay in the net close to the anode remain exposed. It also can be potentially dangerous for people. If the anode handler is using an anode modified as dip net, but forgets to switch off power because the fish are coming fast and furious. When the netter swings the live anode with fish and dumps it into a bucket, some one's may be in or near the bucket, and that person could be shocked if not properly protected. This technique also disrupts the electrical field every time the anode net is removed from the water to transfer fish to a collection bucket. A net on the anode can be helpful, however, in creating currents in deep pools to move the fish around, increasing their contact with the electrical field.

For electrofishing operations with large crews, the supervisor should not directly participate in the fishing. A real-life example illustrates this point. In an operation with a generator floating in a small raft, towed by one crew member, some one else was operating the anodes and still other members were conducting the dip netting. The anode has a net, and the person operating the anode was netting as well. The crew chief was farther out in the streams looking for stray fish and picking up fish on the edge of the field. Yet another crew member was walking along at the stream edge with a bucket into which all the netters were placing the fish. The plastic bale on the bucket handle was missing and the bucket carrier was holding the metal bale with his bare hands. The person operating the anode net came over to dump some fish into the bucket, forgetting to turn off the power, and touched the bucket handle with the anode. The bucket holder was electrified and fell into the water, which was about six inches deep. The dip netter who saw him rushed over to help, with their live anodes still in the water. The crew chief, seeing what was happened, ran to the raft and shut the power off. Although the netters had kill switches on their anodes, they did not think to let go of them. Later when asked if he was conscious and aware of what was happening, the bucket handler replied that he knew perfectly well what was going on, but could not get his arms underneath himself to push up. All he could think about was drowning in a few inches of water.

This event illustrates how personal injuries can occur during electrofishing. However when proper safety precautions are used by trained personnel, the frequency of occurrence of electrofishing accidents is very low.

Standardization

There are three primary sampling strategies, each suitable for different objectives, (1) increasing accuracy by reducing bias, (2) increasing precision by reducing variance, and (3) increasing the specialization or changing the scope of sampling from a general survey to a specialized sampling that targets a particular size group or species.

Many agencies are particularly interested in standardization of electrofishing protocol. The New York Department of Environmental conservation for Aquatic Ecosystem has developed an excellent standard operation procedure manual for Centrarchid sampling (Green, 1989).

One way to improve standardization is to develop tables that relate minimum power, water conductivity and voltage output. Using these graphs, it is possible to ensure that fishing power is maintained under different conditions. To do so, however, requires metering of both output peak voltage and output peak current. These meters are not standard on commercial electrofishing units. However, such meters can be added at additional cost.

The use of surprise tactics (*e.g.* turning power on after entering an area with abundant fish cover) is very effective for improving efficiency. In streams in the mid-west, large fish are very adept at escaping electrofishing capture by using available cover. Failure to effectively shock areas of abundant cover results in a catch skewed toward smaller fish. Another tactic to improve standardization is to surround cover features with block nets.

Testing of Electrofishing Equipment in the Field

Calibrating and testing the performance of electrofishing equipment in the field is an extremely important step in standardizing electrofishing practices and in providing the essential information for accurate data interpretation. The recommended method for calibrating electrofishing equipment is mapping the voltage gradient field. This is done by taking measurements of voltage gradient at varying distances from the anode(s) using either an oscilloscope or peak voltage detector.

A useful piece of field testing equipment is a peak voltage gradient detector. Although the detector does not provide information about the shape of a waveform as does an oscilloscope, it is easier and faster to use in the field. One of the most useful tools is a voltage gradient meter that measures peak voltage. It consists of a probe and detector, which can be plugged into any voltage reading device. A.L.Kolz (National Wildlife Research Center, 3550 East Brook Dr. Fort Collins, CO 80525) has developed the detector, and although it is not produced commercially, he will provide any one who requests it, the specifications for construction. The probe has two exposed metal screws located 5 cm apart. The circuitry automatically divided by five to provide voltage gradient (V/cm). It has two ranges, 1x for 1-5 V/cm and 10x for up to 50 V/cm. It has spring loaded circuit breaker, so it will not stay on accidentally and drain the two 9-volt batteries. Holding the probe parallel to the anode gives the maximum voltage gradient and rotating it 180 degree yields an equal but negative readings. Holding it perpendicular usually corresponds to a lower reading, more in line with the iso-voltage lines.

Taking voltage gradient measurements at the anodes provide information about the maximum power output. Determining the distances from the anodes at which the voltage is equal to 1 V/cm and 0.1 V/cm provides an estimate of the extent of the effective electrofishing range based on studies using gold fish. Although the rule of thumb for effective field strength is 0.1 to 1 V/cm, it is unclear whether a similar general rule applies for fish injury or mortality. One opinion is that voltage gradient does not effect injury rates. According to this opinion, once a fish is held in the field and is reacting to it, it becomes epileptic. This means that the best way to prevent fish injury is to use low-energy waveform. The other opinion is that voltage gradient is a very important factor in injury rates since voltage gradients clearly influence rates of fish stress. The difference in opinions centers on a disagreement of the cause of injury. Insufficient data and lack of peer review on the subject prevents conclusive support of either of these opinions.

An audio amplifier can be used to detect the presence and frequency of electric current rather than risking injury by touching the water with fingers. Information from voltage gradient probes and oscilloscopes can be used to assess waveform and map field characteristics. For example, oscilloscopes can detect leading edge spikes or waveforms. If these spikes are not inherent in the operation of the scope (some scopes are unable to reproduce fast rise waveforms), then the spikes may indicate equipment failures, such as, probe malfunction, presence of additional electrodes or a collapse in the field at some location.

Modifying the array of electrode changes the shape of the field. For example, if the droppers of an umbrella array are spread out more, the overall field strength and shape changes. If the resistance of an electrode system changes, such as, through corrosion build-up on the electrodes field strength is diminished. For these reasons electrodes should be polished periodically.

Electrical field maps for several boat and backpack shockers were evaluated. For the first boat unit a peak voltage gradient of 0.1 V/cm occurred at 0.7 m from the anode. For the second unit, the field extended to 3 m using the same applied voltage setting. This difference in performance has also been seen for backpack units of the same model and manufacturer. Some possible explanation are;

1. Cathode configuration. The more powerful system used the boat as a cathode, while the lower powered system used dropper cables for cathode.
2. Power availability at the anode. There may be differences in the distribution of power in the system. Large cathodes may be able to drive power through a system more efficiently.
3. Anode configuration. When two anodes are located near each other, they may act like a single larger anode.

It would be helpful to store waveform in an oscilloscope, so that two units could be compared to help explain observed voltage gradient differences.

The importance of conductivity and temperature should always be considered. Conductivity is the single most important environmental factor in electrofishing. Conductivity cannot be estimated based on water clarity or color. It is dependent on the area's geological characteristics and anthropogenic inputs. Although conductivity is an important determinant of power density, it does not affect the voltage gradient mapping of a field. In mapping voltage gradient fields, it is important to operate the same electrode configuration with the same applied voltage. Doubling the applied voltage should result in a field of the same dimensions, but with voltage gradient readings twice as high, because the dimensions of the field are determined by the geometries involved. Temperature has a separate biological effect, it affects the ability of the fish to float, respond and escape.

Chapter 3

History of Electrical Fishing

Fishing with electricity has developed from different origins in different environments, and thus has a confusing history. In this it resembles most practical types of activity which depend on the exploitation of a physical principle.

The first definite record is the patent granted to Mr. Isham Baggs, living in Islington, London, in 1863. This is notable in revealing a thoroughly practical approach which must have been the outcome of considerable experience; he advised the use of polarizing spectacles to help the operator. At that date he was handicapped by the lack of any powerful electric apparatus, and relied on batteries. Like many after him, augmented the voltage by using an induction coil, and thus pioneered the exploitation of high powered gear.

Following Isham Bagg's work, there was no practical development of the process, but studies took place predominantly in Germany, on the orientation and movement of fish exposed to electric current (Mach, 1875; Herman. 1885). In England, Loeb and Maxwell (1896) demonstrated these reactions were due to stimulation of the motor neurons and were forced, not voluntary. Blasius and Schweitzer (1893), however discovered the phenomenon of *galvanonarcosis*, a state in which fish appears to sleep with a relaxed body when it is facing the anode. This is in contrast to strong stimulation they observed when the fish was facing the cathode.

It seems that practical electric fishing developed mainly in Germany with Holzer (1932) and the support of physiological research of Scheminzky (1924). However, early in the 20th century there are accounts of Japanese fishermen using induction coils to drive eels out of their burrows into a net (Takahashi, 1895). Also a patent claim by Larssen in 1912 covered the use of electricity to catch a variety of aquatic creatures from fish to seals, suggesting that the claimant recognized the possible potential.

The earliest record of an electric fish screen was in 1917 in the United States when Burkey took out the first (Patent No. 1269380) of many patents. Systematic work on screens, however, really dates from the masterly analysis of Mac Millan (1928) and Tauti (1931) which enunciated the principles of aquatic fields and their logical use as fish screens.

After the last war, the development of electric fishing and electric screens continued in Germany, primarily with Denzer (1956) and Halsband (1959) as the driving forces. At the same time electric fishing engines were being produced commercially and this opened up a whole new dimension. During this time the control of fish behavior was quite familiar in Germany by direct current in tanks and in America by alternating current, and a series of papers appeared, mainly in Germany (namely, Scheminzky, 1922, 1924, 1934; Scheminzky and Kollesperger, 1938).

Post-war American was initially concerned itself with the construction of portable apparatus capable of being used by a single individual and suitable for locations difficult to access. More recently, however, design shifted towards large fishing devices based on what can only be considered as fish-screens carried on a special rafts moved by propulsive motors in open water, The formerly ubiquitous employment of alternating current was replaced by pulsed direct current provided by electronic devices from an alternating supply. This had an effect on the thinking underlying the whole process, introducing a degree of formality into the procedure, which was previously unnecessary, as a precaution against fatal accidents which had by now occurred. The days of experimental apparatus has passed. In general, fishing machines had to be commercially constructed to approved standards.

The situation in Great Britain after 1945 developed from both German and American routes. The fishery authority of eastern England, where the rivers were mainly sluggish and inhabited by coarse fish, followed along American lines and used alternating current. Simultaneously, in the fast flowing, high conductivity chalk streams of southern England, experiments took place with direct current to facilitate the removal of coarse fish from these trout waters.

During the time considerable research and development took place elsewhere. For instance, in the Soviet Union, Strakhov and Nusenbaum (1959) were developing electric screens and Shentiakov (1960) electric trawling in lakes. In North America, Applegate *et al.* (1952) were producing an efficient electric screen against the invasion of sea lampreys into the Great Lakes and in New Zealand, Burnet (1959) was experimenting with electric fishing.

In 1957, in Hamburg, a first synthesis of electric fishing problems was resolved in the first FAO International Fishing Gear Congress in 1963, although discussion on electric fishing was minimum.Thereafter Blanchteau *et. al.* (1961) in France and Nusenbaum and Faleeva (1961) in the Soviet Union, introduced the elements of comprehensive theory of electric fishing, based on physiological principles.

In 1965, the European Inland Fisheries Advisory Committee (EIFAC) arranged a meeting in Bairritz, of workers in the electric fishing field. The working party received papers on various aspects of the technique, which was subsequently embodied in a book *Fishing with Electricity* (Vibert, 1967 a) representing its report to the Fourth Session of EIFAC at Belgrade in 1966.

In 1973, the working party reconvened in Gysisko, Poland, to compare the efficiency of portable apparatus used by participants at that time (EIFAC, 1973), and elaborate tests on the fatigue and mortality induced by the gear. The results were published by Chmielewsky (1973 a,b). During this period, a book on the general principles of electric fishing was produced by Sternin *et. al.* (1972) and Halsband and Halsband (1984) updated their book on the subject.

More recent developments were reported to the EIFAC Symposium on fish population studies held at Aviemore, Scotland in 1974. This provided descriptions of new designs in fishing apparatus, and the instruments for the detection and counting of migrant fish passing a census point in a channel. In 1975, Lamarque, in conjunction with the Food and Agriculture Organization (FAO) completed a comprehensive study on electric fishing in tropical waters which elaborated on the theory of Blanchteau *et. al.* (1961), allowing the choice of the best current for catching fish and opening up the practice of fishing in sea water and low conductivity water (Lamarque, 1977 b).

At the University of Wisconsin, Novotny and Priegel (1974) developed a boat-mounted, multi-anode, boom-array electric fishing system, which was the forebearer of today's commercially available electrofishing units. In the past most electric fishing equipment was "home made" by biologists with limited electrical engineering backgrounds. For today's fisheries biologist, there exists a wide inventory of field tested, dependable, commercially available electric-fishing equipment. The developments in today's electric-fishing gear have been directed towards reducing harm to fish, saving time and money, reducing the possibility of serious injury to fishery workers, and increasing catch per unit effort.

Under present conditions, technical advances in electronics have now made it possible to design and construct light-weight fishing apparatus with virtually any current characteristic, thus increasing the potential for future development.

Chapter 4

Electrical Fishing Technique Adopted in Different Regions

The use of electricity for stream population studies seems to have originate in eastern Germany. Shetter (1938) translated and reviewed a short article from a German fisheries publication which described the electrical equipment needed. This article also listed the standing crop of trout and its value, for certain Bavarian trout streams in three different years. Biologists and technicians connected with the New York State Conservation Department assembled and employed the shocker in that state first in 1939 (Haskell, 1939).

In Michigan State, the shocker was first operated on Hunt Creek in mid-summer of 1942. The unit has received much use since then, and there was as increasing demand, both from the biological staff and the hatchery personnel, for the services of this type of gear, an additional unit of higher wattage was acquired and placed in use.

The source of power is a one h.p. gasoline motor which drives a generator of 500 watts capacity. The normal current output varied between 1.5 and 2.5 amperes, depending upon the conductivity of the water and the distance between the electrodes. The motor and generator were mounted on a carrying frame which can be transported by two men, or rested in the bottom of a boat. The electric current was conducted to the water through a rubber covered two-wire cable to a pair of electrodes which were attached to wooden handles. As each electrode was connected to one pole of the generator, this forced the electric current to use the water as a pathway in completing the circuit when the electrodes were submerged in the water. Depending on the shape and size of the electrodes and the depth of the water, the electrified field appeared to be similar in shape (when viewed from above, and judging by the actions of the fish) to the conventional diagrams of fields of attraction around two magnetic poles. Any fish which entered the electrified field was momentarily stunned and was picked up with a scap net. Apparently the rapid reversal of the current flow in an AC circuit affected the nervous system. The larger fish succumbed readily to the shocker, but the electrodes need be close together to knock out small fish (fingerlings and fry) since their smaller bodies did not absorb as great portion of the electric current as did the larger fish.

In field operations the captured fish were removed by scap net to fresh uncharged water. Fishes recovered from 30 seconds to five minutes, depending upon the size of the fish and severity of the shock. Less than one per cent of the fish handled were killed. Only when a fish remained too long in the charged field, or came to rest unobserved against an electrode was burned mortally.

Passage of Current through Water

Electric current passed through a 2.5 ampere fuse or circuit breaker which protected the generator from an overload. From this point it passed through an ammeter, if desired, thence through a choke coil and fixed condenser to an electrode. When both electrodes were in the water, the electric current flowed through the water to the other electrode and back to the generator to complete the circuit.

Modifications of electrodes were numerous. Hager (1934), the German experimenter, used 10-inch square copper plates on wooden handles to introduce the current to the water. The New York State operators drove a single stake or series of stakes into the stream bottom for one electrode, and floated a metal screen connected to the opposite pole of the generator for the other electrode. They noticed that this reduced the voltage available in the water for shocking fish in too many types of stream bottom soils, and instead of stakes driven in the bottom substituted a bare copper wire. However, trials in Michigan with the later electrode set up proved unwieldy in snaggy streams, and the floating screen created a riffle which made for poor visibility of the fish shocked.

Development of Electrodes

Electrodes were made from ½ inch mesh hardware cloth soldered to a foot – shaped frame of ½ inch reinforcing steel, and were about 16 inches long and 6 inches high at the heel. While these gave satisfactory results, they could not be operated much farther apart than 8 feet, and seem to burn a number of large fish. Noting this difficulty, new electrodes were build, which was, in reality a widely spaced induction coil of bare wire on a wooden frame 14x6x1 inches. About 55 feet of copper or phosphor-bronze wire (# 14, 16, or 18) were used. With this modified electrode it was possible to shock all sizes of fish passing between the electrodes when they were as much as 12 to 15 feet apart and in water of 2 ½ feet deep. Thereafter, triangular shaped electrodes on which the wiring of induction coils was wound in a plane parallel to the stream bottom rather than vertical to it. This carried approximately 95 feet of wire on a triangular frame which is 2 feet high with a 22 inch base. The wire was spaced at 3/8 inch intervals. This model had worked the best of any other tried in 2 ½ feet deep water. In water deeper than this, the triangular model tend to cause the generator overwork and heat up. Also, it did not gave a large enough charged field, thus allowing many fish to escape either on the bottom or near the surface. To offset this, a long narrow type of electrode was built (36x6x1 inch) and wired with 55 feet of wire spaces 1 ½ inches apart. Trials with this deep water electrodes indicated that it will shock more of the fish from pools 4 to 6 feet deep. With this latest type of electrode the charged field seems to be narrower and deeper.

Operation of the Shocker

In actual stream-side use, the motor was started with no load on the generator, and after a short warm-up period the electrodes were plugged and submerged. The customary procedure is to work upstream against the current. In this manner, the crew was not rolling the water to be shocked, and the stunned fish were carried to them by the water current. Although the equipment can be operated by two individuals, the ideal crew size for intensive work is six men. In population study operations each electrode is handled by one man. The electrodes carriers wade on the outside of the area to be shocked, and usually between and slightly behind them one or two assistants were stationed to net the shocked fish as rapidly as they are seen. A fifth individual carried a pail of fresh water to receive the fish or pilled the boat floating the shocker and kept the connecting wire free of snags. The sixth member of the party tallied the fish handled and recorded measurement, tag numbers or missing fins of marked fish recovered or any other data required.

In conducting population studies in a section of a stream, two complete trips were made. On the first trip, all fish captured were given a distinctive mark before they are released, such as clipping a small piece from the tail fin. Careful record of the number of fish caught and so marked was kept. Two or three days later the shocking crew went to the same water and a record was made of the number of marked and unmarked trout captured on the second trip. From these records on the number of marked fish in the stream and the number of marked and unmarked fish taken in the second check, the total population of the stream can be computed.

Electric Seine

Widespread use of the electric shocker for making fish population studies dates from Haskell (1940). Since that time the method has come into rather common use by fishery research agencies and many improvements have been made on the equipments and techniques as originally described.

Limitations of the Method

The great advantage of the method, that of providing a means of sampling fish populations in waters where bank and bottom conditions make seining impossible, is overbalanced in the minds of many, by its limitations. Some of these limitations are;

1. Equipment is cumbersome and costly - The equipment commonly used consists of a more or less elaborate electrode hookup to be handled in water, several hundred meters of cable to transmit the current, voltmeters and ammeters to measure the electric energy, a more or less complicated transformer or rheostat arrangement to control it, and a generator to produce current. The total cost of these items amounts to lots of money and most of them are difficult to obtain.

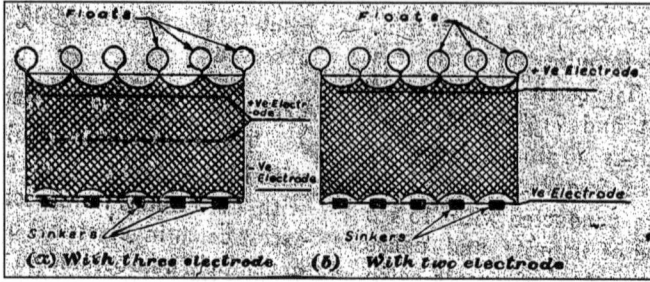

Figure 3: Electric Seine Developed by Dr. K.P. Biswas with Three and Two Electrodes.

2. The following facts are evident in handling the equipments. Even the so-called "portable" power plants, if they are of larger and more powerful types, are extremely heavy and difficult to handle on rough and muddy stream banks. The meters used to measure electric energy are delicate instruments requiring careful handling. During operation they and the rheostat require constant attention from a member of the crew. Variations in the resistance within the field makes necessary changes in the rheostat setting to prevent the amperage from rising above the limits lethal to fish. Generally such hookups are protected by fuses which burn out before the established limit is exceeded. These precautionary arrangement may be the cause of many irritating delays. Several hundred meters of electric cable can, also cause many delays as it becomes lodged or snarled while being dragged along rocky stream beds or brushy banks. Finally, the electrode hookup, although varying greatly in design, will usually require at least two men to operate it, will have wires which frequently become snarled or hung up, and will have numerous connections, that require considerable time to make and release when starting and stopping the work and frequently become disconnected during operation, also causing delays.

3. Method effective only in narrow, shallow and clear stream - The method has been used most in northern trout brooks and the equipment developed has been designed for use in streams of this type. Little seems to have been done to adapt it to wider, deeper, more turbid waters, even though its use in wide streams or in lakes would be most valuable. The assumption that it would be impossible to recover stunned fish in deep turbid water has been generally accepted.

4. Effectiveness of the method varies greatly with chemical content of water and soil - This statement undoubtedly was true of the device originally described by Haskell. Improvements described by Haskell and Zilliox did much to overcome the difficulties caused by soil resistance although the device had other limitations which were mentioned.

5. Method is dangerous - Haskell and Zilliox took great care to point out the possible danger involved in operating an electric-shocking device. Their warning was so urgent that it may have served to deter some who otherwise would have been interested in trying out the method.

These are a few of the more common limitations of the method as it was ordinarily used. In 1945 it was decided to try out an electric-shocking device (commonly referred to as the "electric seine") as a method of collecting in fishery-census work in Missouri. Several changes in the conventional type of apparatus have been made and several new techniques have been developed. All new developments have been tried out under field conditions and only those which were practical under such conditions have survived.

Equipment

A hundred ten volt, 60-cycle alternating current generator of 500 watts capacity had been found to provide a satisfactory source of power. Experiments had shown higher voltages and greater watt capacities to be unnecessary. In the interest of portability, the motor be of air-cooled type and to have a plant with casting of light metal alloys rather than the cast iron, even though the cost may be somewhat more. A plant equipped with an automatic cut-off which prevents overloading the motor when the rated capacity is exceeded, makes fuses unnecessary and is a distinct advantage. Such plants weighing less than 75 kg and equipped with carrying handles were available then.

The electric hookup used was simple, easily constructed and was easy to operate. Eight floating grids connected to one line from the power plant, form one group of electrodes. The floating grids were 5 cm to 7.5 cm pieces of white pine of one meter long, covered on three sides with 1.2 cm mesh hardware cloth. The wire connecting them was interlaced with the hardware cloth for several centimeters at each end and is held in place with staples. These grids were arranged in line 3 m apart from center to center. Connected to the other line from the power plant were seven drag wires- 2 m lengths of flexible, un-insulated wire cable weighted at the outer end with lead seine weights to keep them on the bottom. The power line with these drag wires attached was arranged so that the wire hang suspended midway between each pair of floating grids. At each end of electrode hookup was a probe pole, a light wooden pole 2.5 m long, the lower 0.5 m of which was covered with metal, preferably copper. This pole was connected by means of a 3.5 m length of insulated wire to the power line to which the drag wires were attached. These, with the drag wires, form the opposite group of electrodes. The whole apparatus was simply a number of resistances connected in parallel, the water between the various pairs of electrodes forming the resistance. The two power wires were fastened together with friction tape at intervals so that the floating grids kept both near the surface. The size of the floating grids was just sufficient to keep the wires afloat, thus helping to keep the bulk of the entire device at a minimum. All connections were permanent except that of the power lines to the generator. This arrangement eliminated delays in starting and stopping work and during operation. Wire for the power lines should be flexible, insulated copper cable, size No. 10 or No 12. Good quality cable will give less trouble of breaks and short circuits. Meters to measure the current and transformer or rheostat were not needed after the potential of the device was established. With a 110 volt, 500 watt power plant equipped with automatic cut-off, little more than 5 amperes of current ever went through the lines. This amount was sufficient to stun most fish but not enough to cause serious injury. They recovered in a few minutes when removed from the electrical field. A long transmission line is not necessary if the power plant is placed in a boat and floated behind the shocking device during operation. With this arrangement a line of 10 meter or less in length may be used and delays from snarls *etc.* were practically eliminated.

Operation

Two men were required to operate the electrode hookup in the water. A length of light rope attached to the outermost floating grid is tied around the waist of each operator, leaving his hands free to manipulate the probe pole and a dip net for picking stunned fish. One or more additional men devote their entire attention to the recovery of fish. One man handled the boat and power plant and assist in the recovery of fish. Equipment for storing fish *etc.*, can be carried in the boat.

Ordinarily, a stream section is worked by starting at the upstream end and working downstream. This procedure was followed for several reasons; the natural tendency of fish is to swim upstream, sediment stirred up by the operations is carried on by the current, and the equipment was more easily handled. With such a device, streams up to 35 meter wide can be covered thoroughly in one trip. On narrower streams the device is simply allowed to bag as much as it will, the only effect being to put more of the total water area within the electric field. Wider streams can be worked by following the shoreline with one end of the apparatus while the other end is extended full length, and the whole device is kept at an acute angle with the bank. By moving along both margins in this way a fairly thorough sampling of the fish population can be obtained. If the middle portion of the stream is too deep for wading, the outside man may be carried in the boat, although this technique as a rule is not especially effective, if for no other reason than that the stunned fish were difficult to recover in water so deep.

This equipment and these procedure go far toward overcoming the first two limitations mentioned. The third, namely that the effectiveness of the method varies with the chemical content of water and soil, has not proven to be true with the apparatus described. Soil resistance is unimportant since with this device very little of electric current goes through the soil. The records showed that the apparatus had been used with apparently equal success in waters of which the calcium carbonate content varied from 75 to 177 p.p.m. It seems that most natural waters, at least running water in streams, contain enough dissolved minerals to transmit electric current readily. The final limitation, that the method is dangerous, was best met by following the admonition of Haskell and Zilliox; " A major precaution for the safety of the crew is proper supervision". All crew members are required to wear waders and electrician's rubber gloves at any time the power plant is in operation. So far, the greatest casualties had been minor shocks occasioned by falling or overtopping waders while within the field.

Tests of Efficiency

Haskell and Zilliox described a number of tests of the efficiency of the electric-shock method. Their work was all conducted on small trout streams. They concluded that for trout more than 10 cm long recovery was 80 per cent or greater and that for fish of this size width of stream, depth of water, type of section and type of bottom do not have an important effect upon recovery. They found the recovery of fingerlings to be highly variable. Schuck (1946) found that in the areas he studied, the size of the trout was the most important factor affecting efficiency. His recovery figures ranged from 95 per cent for age-groups III to IV (21.4 to 41.2 cm) to 50 per cent from 0 age-group (5.2 to 10.5 cm). The larger trout were captured more efficiently, possibly they were more conspicuous after being stunned.

With the electrode hookup described, there was no place where a fish passing through the field could be more than 1.3 m from an electrode. At this distance most fish of moderate or large size were stunned, but small fish were freely observed to swim through the field unaffected. They must come to within 0.3 to 0.6 m of an electrode to be stunned.

The physical factors which was observed to influence the efficiency most were, in order of importance; turbidity, depth, type of bottom and velocity of current.

Experiments to test the efficiency of the shock method were started during 1947. They were not complete and it appears that factors other than the efficiency of the seine may have interfered to cause unexpectedly very low recovery figures. The experiments all followed the same general pattern.

A section of stream was blocked off by means of 6 mm mesh seines placed at each end. The blocked-off area was stocked with fish of as great a variety of species and size as could be obtained. Each planted fish was weighed, measured and marked, either with a tag or by having a fin clipped. It is recognized, of course, that fish so planted might be more easily recovered than undisturbed fish native to the area. The fish were given short time to become acclimated, then the section was covered as thoroughly as possible with the shocking device. The number of times it was gone over depended upon such factors as the time required for the resulting turbidity to subside or the length of time it was practical to maintain the blocking seines. In tabulating the results, fish of various species were dived arbitrarily into size groups by total length. For sunfish the size range for each group was 2.5 cm; for black bass, redhorse and fish of similar size, 5 cm; for carps 10 cm.

It would seem to be evident that the efficiency of the electric-seine method, like that of any other method of collecting fish, varies with the physical characteristics of the area. Each stream section presented an individual problem. Some additional facts were also evident. In most places the percentage of large fish recovered was greater than that of the smaller individuals of the same species. This fact can be explained in two ways; small fish were more likely to be overlooked, and small fish required a greater amount of electric current to stun them. It was also evident that most species of fish were affected. Finally it should be evident that large numbers of fish can be collected by this electric shock method from areas in which it would otherwise be impossible to sample the fish population.

Chapter 5

Electrofishing Boat

The use of electrofishing equipment mounted in flotilla to sample fish populations in freshwater lakes and rivers has been an accepted and useful practice for many years. The equipment used is nearly always locally designed and great variations in techniques and performance were reported (Myers, 1951; Rollefson, 1958; Burnet, 1959; and Patten and Gillaspie, 1966). To a large extent this wide variation in equipment and technique is a result of great variability of water conditions, fish species and sampling needs in freshwater lakes and large rivers coupled with the fact that almost any electrofishing system will allow capture of some fish under conditions suitable to the particular system. In contrast, the less varied demands of electrofishing in small streams had resulted in more uniform and thus more highly developed methods (Cuinat, 1967; Novotny and Priegel, 1971).

With the objectives to survey the equipment and techniques in use, to clarify the impact of the electrical variables influencing performance, to develop guidelines for constructing and operating fixed electrode electrofishing boats, and to pinpoint areas where additional developmental research should be carried out to provide future improvements in electrofishing methods, Department of Natural Resources, Madison, Wisconsin, had undertaken field study with the leadership of Donald W. Novotny and Gordon R. Priegel in 1974.

Because of the wide-ranging objectives the approach used in carrying out the study involved a number of separate but interrelated activities, including: (1) literature searches on electrofishing systems and the response of fish to electrical stimulation; (2) field tests and observations of existing electrofishing boats; (3) theoretical and experimental evaluation of electrode systems; (4) design, construction and field testing of an experimental boat capable of operating with alternating current; and (5) surveys of performance of existing, modified and experimental boats over a wide range of operating conditions.

The results of these various activities were organized into two categories; basic concepts and design guidelines and experimental and operational systems.

Problem Areas

Basically, the function of an electrofishing system is to produce sufficient electrical stimulus in fish near the electrodes to permit easy capture by netting. In an alternating current (AC) system the electrical stimulus simply immobilizes fish (electronarcosis or electrotetanus) requiring the net handlers to dip fish from considerable depths or distances from the boat. Often fish were immobilized at depths beyond reach of the dipper and were not susceptible to capture at all. In direct current (DC) or pulsed current system, fish near the anode exhibited forced swimming (electrotaxis) toward the anode and hence more easily netted. It is quite well established that fish were much more susceptible to AC than to DC and hence the radius of action of an AC electrofishing system operating at a particular voltage is much greater than a DC system at the same voltage. Pulsed DC is more effective than DC and to some extent combines the desirable forced swimming response of DC with the larger radius of action associated with AC.

Work done on electric shock response in humans (Dalziel and Lee, 1968) had shown that electric current is the best measure of the strength of sensible shock effect. Similar results regarding electrofishing effects were indicated by the work of Cuinat (1967) and by the present and earlier work on stream shocking (Novotny and Priegel,1971). These results imply that the primary function of the electrofishing system is to establish an electric current in the water near the fish. A portion of this current (depending upon the ratio of fish and water conductivity) will pass through the fish and if of sufficient magnitude will elicit the desired response. The important point is that the current is the variable of concern. Electrode voltage, spacing, size *etc.*, are important parameters only to the extent that they influence the current.

Based on this concept, for any electrofishing system there exists some minimum value of current per electrode which will affect fish out to some specified distance from the electrode. The minimum current will depend upon electrode shape, nature of current (AC, DC, or pulsed DC), desired range of action, fish species, water temperature, water conductivity and many less obvious factors. Of these factors only electrode shape and type of current are under control of the designer. All other factors are external variables which must be compensated for by variations in the operation of the boat. Based upon the surveys, field tests and literature review carried during the study, the following specific problem areas were identified.

1. Range limitation – The distance at which fish are affected can be too small to produce useful sampling results. This is often caused by incorrect electrode arrangements and insufficient current (and power).

2. Water conductivity – Low water conductivity makes it difficult to attain sufficient currents to produce useful electrofishing responses. Lenon and Parker (1958) who found extreme conductivities in Appalachain mountain streams, attacked this problem by adding salt to the water to improve

electrofishing. Extremely high water conductivities call for currents too large to be supplied by portable equipment without special electrical control methods. For very high conductivity, DC is ineffective. Other factors that are important are voltage and electrode size.

3. Water clarity and vegetation — Turbid water and excessive vegetation restrict visibility and reduce the value of the immobilizing capabilities of the AC electrofishing. Pulsed DC and DC offer potential solutions to this problem.

4. Water depth — Fish which are immobilized at depths exceeding 0.9 to 1.2 m are very difficult to capture by ordinary netting procedures. Pulsed DC and DC should overcome this problem.

5. Bottom materials — High conductivity bottom materials tend to "short circuit" the current out of the water into the bottom material reducing the electrofishing effectiveness and sometimes overloading the power source. Little can be done to remedy this situation except to use short electrodes, near the surface to avoid direct contact between electrodes and the bottom.

6. Water temperature — The conductivity of water and water temperatures are more or less a straight line relationship; that is, as the temperature of the water increases, the conductivity of the water increases at an equivalent rate (Sigler, 1969). Theoretically, success of electrofishing should increase with temperature, but workers in the field have noted differently. Smith and Elson (1950) believed that salmon parr exhibited the best response below 25 degree Celsius and suckers at less than 20 degree Celsius. Webster, Forney, Gibbs, Severns and Van Woert (1955) had greater success in shocking brown trout at 7.8 degree Celsius than at 16.7 degree Celsius when both AC and DC were used. The response of brook trout to DC current is low in cold water, but increases with the temperature to 10 degree Celsius. As the temperature increases further, the response decreases (Elson, 1942). Most salmonids are more easily captured by electrofishing when the water temperatures were low, 0 to 10 degree Celsius (Vincent,1971).

7. Fish mortality — Mortalities caused by AC electrofishing are higher than those caused by DC or pulsed DC (Taylor *et al.*, 1957) and the gross physical damage from AC can be severe (Hauck, 1949). Harmful effects from pulsed DC are usually a result of excessive exposure or intense electrical field (Pugh, 1962). It can be easily demonstrated that fish can be killed with electricity in their natural environment, as well as under laboratory conditions (Pratt, 1954; Godfrey, 1956).

With electrofishing boats, mortality is usually associated with regions of excessively high current density close to the electrodes. Potential solutions involve electrode designs which avoid such regions of high current density by using larger electrodes and lower voltages. No injury to fish has been observed with DC or pulsed DC.

8. Fish size — Individual variation is notable among fish even though they are of the same species and have similar lengths. The laboratory experiments of Haskell *et al.* (1954) on brown trout demonstrated this variability.

   The larger the individual of a species, the more sensitive it is to a given electric shock (Mc Millan, 1928; Mc Lain and Nielsen, 1953; Taylor *et al.*,1957). Fish absorb power as a function of body surface area and particularly length (Holzer, 1931). Also, the greater resistance of smaller salmonids and possibly small fish of other species as well, further reduces their response to shocks (Nakatani, 1954). All these factors contribute to capturing a greater proportion of large fish than were actually present in the population.

9. Fish species — Some species are quite difficult to capture by electrofishing. The most notable are northern pike and muskellunge. It appears likely that this is caused by the strong swimming ability and natural tendency to escape danger by rapid swimming coupled with high sensitivity to electric fields. Operating at higher boat speeds when seeking northern pike and muskellunge is a potential advantage.

   Large differences in response have been observed between various fish species when pulsed DC was used. Some selectivity in electrofishing may be possible because of the differences. There may be optimal wave forms and pulse rates which will produce selective and efficient electrofishing for various species.

10. Equipment and operating problems — Numerous problems associated with equipment limitations and lack of operating guidelines exist. Examples include inadequate lighting systems, insufficient power, lack of voltage control, lack of proper instrumentation, poor electrode design *etc.* The problems identified demonstrate the need for developing guidelines for electrofishing systems capable of adjusting for variations in external conditions.

General Design

Although the design and operation of effective electrofishing boats is still to some extent an art because of lack of basic data on fish responses to various levels and types of electrical stimulation, there are a number of basic concepts which are useful in guiding the development of new boats.

Components of an Electrofishing Systems

The components of a fixed electrode electrofishing system can be classified into four subsystems according to function. These are;

1. Boat and mechanical – to effectively and conveniently carry the complete system.

2. Electrical - to generate, control and deliver the electrical energy to the electrodes,

3. Safety - to provide proper safeguards for operating personnel, and

4. Electrodes – to properly couple the electrical energy to the water.

Boat and Mechanical Components

Selection of the boat, motor and other mechanical components is largely a matter of individual choice relating to convenience, availability, ease of maintenance, *etc.* Sixteen to eighteen feet square ended aluminum boats was found to be satisfactory. An aluminum boat offers the advantage of simple reliable grounding of all electrical equipment through the physical attachment of the equipment to the boat. Addition of any metallic structural parts to a fiberglass or wooden boat would require careful grounding practices to assure electrical safety, whereas such structures are immediately grounded through structural members in an aluminum boat.

Electrical System

In addition to the main generator the electrical system of an electrofishing boat should include an auxiliary generator and battery, metering, disconnect switches and overload protection, voltage controls and power conversion equipment, and a safe and reliable system of interconnecting the various components. A means of rapidly disconnecting the power in case of emergency must also be provided as a part of the safety system.

Main Generator

A three-phase AC generator provides the greatest flexibility. In AC electrofishing all three phases can be effectively utilized by proper electrode design. If DC or pulsed DC electrofishing is desirable, a transformer rectifier system provides a voltage controlled DC source without need for large filters as would be required with a single phase generator. Compared to a DC generator, the ease of controlling voltage with transformers to match the generator to water conductivity conditions is a major advantage of an AC generator.

The frequency of the generator (cycles/second or Hz) does not appear to be a critical factor in AC electrofishing effectiveness 60 Hz, 180 Hz, and 400 Hz have been used without significantly different effects. Since higher frequencies offer weight advantages in designing suitable transformers, the highest of the commercially available frequencies is preferred.

Power rating is determined by the maximum water conductivity to be fished, the size of electrode which can be supported by the boat and by weight limitations. In general, electrofishing range and hence effectiveness increases with increased power and it would therefore appear that selecting the largest generator meeting weight limitations is desirable unless only very low conductivity waters are to be fished. Power and voltage requirements for the boats developed and used during the study are given below.

Voltage Ranges for AC Electrofishing

Water Conductivity (micromhos/cm)	Voltage (V)	Power (kW)
10–20	460*	2.0**
20–40	460*	3.5**
40–60	460-390	4.5*
60–100	390-320	4.5*
100–160	320-230	4.5*
160–350	230-160	4.5*
350–700	160-100	4.5*

* Denotes limit imposed by equipment rating; ** Denotes limit imposed by maximum electrode size.

Auxiliary Generator and Battery

Use of an automotive-type, 12 volt DC system for auxiliary power has proved very effective. This scheme offers auxiliary power at all times, independent of whether the main generator is operating, and automatic recharging whenever the main generator is operated. The entire system consisting of generator, voltage regulator, meters and battery can be integrated with the main generator to form a compact self-contained unit. The low voltage is very desirable for safety, the system is reliable and repair parts and accessories are readily available from automotive suppliers.

Meters

To permit proper utilization of the equipment, provision of a suitable set of meters is essential. For AC electrofishing, one voltmeter and a set of ammeters are sufficient. When transformers are used, it is convenient to place the ammeters in the primary circuit so the operator has only one maximum current limitation to remember. In addition to a voltmeter in the primary circuit to monitor generator performance, a second voltmeter in the secondary is useful to indicate the actual output voltage.

For DC operation a voltmeter and an ammeter are essential. In the case of pulsed DC, two voltmeters are useful; one to read peak voltage and one for average voltage. Comparison of the two readings yields information on the duty cycle of the output pulses.

Disconnection and Overload Protection

As in any electrical system a disconnect switch and overload protection must be provided. In an electrofishing system the electrical load can vary considerably because of the changes in water conductivity (caused, for example, by an inlet stream). Such changes can cause overload conditions with subsequent circuit interruption unless the system is comprised by operating well below rated load for the normal water conductivity. For this reason circuit breakers are much preferred over fuses for protection of the main power circuits. Protection for the auxiliary power circuits and for diodes and other components in power conversion equipment can be provided by fuses since the primary need is for protection against misuse or equipment failure.

Voltage Control

The use of transformers to control the output voltage and thus provide a proper match between electrode resistance and the generator characteristic is extremely valuable. This is particularly true when it is necessary to operate over a wide range of water conductivity. As a general guide to the voltage range needed for various water conductivities, the voltage levels found useful for AC electrofishing is presented above. Since the voltage level required depends greatly on the size and type of electrode system, the data presented above should only be considered as a general guideline for electrofishing boats similar in size and type to those used in the study.

The power values given in the above table are included to indicate the levels employed in the study. It may be noted that except for very low conductivities, the power was limited by equipment ratings and not by electrofishing considerations.

The transformers used during the study had fixed range settings as opposed to being continuously variable. This poses no limitation since the main engine-generator could be controlled over a speed range which allowed intermediate voltages when necessary. This arrangement has a considerable weight advantage over a continuously variable transformer since the windings can be designed to be operated at nearly maximum utilization on all range settings.

Interconnection System

Considerable benefit can be gained by combining as much of the electrical system as possible into a single package built around the main engine-generator. This avoids the need for mounting many separate components on the boat itself and results in simpler and more reliable construction.

It is extremely important to pay careful attention to the selection of electrical hardware for use under the demanding conditions of electrofishing boats. Use of marine grade components wherever possible will prove invaluable in eliminating malfunction due to corrosion. Items, such as, power connectors and switches must be carefully chosen.

Safety System

The importance of safety considerations in electrofishing cannot be over-emphasized. The question of safety should be given great attention from both the design and operational point of view. In general, the following items should be carefully considered in establishing design and operational guidelines for safe operation.

1. Equipment must be selected and designed to avoid potentially hazardous situationsAs specific examples, power connectors having metallic parts which could become energized through insulation failures and net handles

which could form an electrical connection between the water and an operator must be avoided.

2. A low voltage system should be used for all electrical functions except the actual electrofishing function.

3. Proper grounding practices to assure that all metallic parts in the boat are joined electrically are absolutely essential.

4. A rapid disconnect system is necessary. In addition, to insure that each operator is at his assigned station, a set of weight actuated switches can be employed. Additional protection can be achieved by using a latching system requiring the crew chief to actuate a start button after observing that each crew member is in his proper position.

5. Crews must be aware of the hazards associated with electrofishing equipment and be properly trained in safety and rescue procedures.

Electrode Systems

The electrode systems employed in most electrofishing boats vary considerably from boat to boat. This causes difficulty in interpreting survey data since the various systems are not easily compared in terms of readily measured variables. To provide a basis of comparison of various systems and to develop design procedures for effective electrofishing electrodes, a considerable portion of the study was devoted to electrode phenomena. This work involved a combination of theoretical, experimental and operational activity.

Electrode Requirements

The requirements of an effective electrofishing electrode system include;

1. Establishment of an effective electric current distribution in the water to be sampled;

2. Avoidance of local regions of unnecessarily large current densities which waste power and are potentially harmful to fish;

3. Adjustability to meet changes in water conductivity;

4. Ability to negotiate weeds and obstructions;

5. Ease of assembly and disassembly;

6. Avoidance of unnecessary physical disturbance to water to permit easy visual observation of fish.

Most electrode designs appear to have been greatly influenced by the last three factors without sufficient consideration being given to the basic question of producing an efficient and effective electrofishing current distribution.

Basic Electrode Phenomena

As the current supplied to an electrode passes into the water it spreads in all directions and creates the electrofishing field. The current and voltage associated with the electrode become distributed through the water and the nature of this distribution is critical to effective electrofishing. The electrical parameters used to describe the distribution are the current density (A/cm2) and the voltage gradient (V/cm). These quantities are measures of the strength of the current and voltage difference at a point in the distribution and are related to each other by the electrical conductivity of the water (micromhos/cm) according to the relation;

Current density = voltage gradient x conductivity.

For a given conductivity, current density and voltage gradient are directly proportional and that increased conductivity results in increased current for a fixed voltage gradient. While current density is easy to interpret, voltage gradient is somewhat less obvious. For the purpose of discussion it can be interpreted as the voltage difference between two closely spaced points divided by the distance between the two points. Summing up the voltage gradient along any path between two electrodes yields the voltage difference between the two electrodes.

There are many references which suggest that current is the variable most directly related to electrical effects on fish. For this reason, the current density in the water is the best measure of electrofishing effects and required to know how the current density distribution varies for different electrodes. As the fish usually have a different conductivity than water, the current density in a fish is different than that which would exist at the location if the fish were absent. This is an important effect, and it is appropriate to consider the current density in water alone as a basic measure of electrofishing effectiveness. The discussion of electrode phenomenon is therefore based on the assumption that there exist definite values of current density in fish which will evoke the various desirable (and undesirable) responses associated with electrofishing. There is some minimum current density at which fish perceive the electrical field (perception level), a higher level at which the desirable effect of electrotaxis for DC or electrotetanus for AC occurs (effective level) and a still higher level at which the undesirable effects of electrotetanus for DC or lethal effects for AC occur (danger level).

As the current spreads into the water around any electrode, the current density must correspondingly drop off steadily as the distance from the electrode increases, Every electrode therefore has a near zone of high current density which may be above the danger level. Surrounding this nearest zone is a region in which the current density is in the effective range. The current density gradually drops off until it is no longer large enough to cause the desirable reactions, and this marks the end of what might be called the effective zone. Beyond this distance the current density continues to drop until it is no longer even above the perception level. Clearly an effective electrode is one which has a very small (or preferably no) danger zone, a maximum effective zone and the smallest possible perception zone.

The way in which the current and voltage are distributed in the water near an electrode (even without fish present) is a complex question. The distribution is completely determined by the size, shape and spacing of the electrodes and it is the complex shape of most practical electrodes which makes the problem difficult. There are two simple shapes, the sphere and the cylinder, which essentially form bounds between which most practical electrodes exist and which permit a somewhat simplified description of actual electrofishing distributions. By examining the behavior of these two simple electrodes and recognizing that most practical electrodes fall somewhere between, it is possible to obtain a basic understanding of electrode behavior.

Summarizing the important properties of the current and voltage distributions of spherical and cylindrical electrodes, these properties are easy to obtain because of the symmetry of the distributions resulting from the simple shape of the electrode. Thus for spherical electrodes spaced more than several radial distances apart, the current spreads essentially uniformly in all directions and the current density drops off in proportion to the square of the distance from the center of the sphere. For cylindrical electrodes, the current density is uniform along the length of the cylinder and hence falls off more slowly in proportion to the first power of distance. As a consequences of these differences, the voltage distribution and hence the resistance of the two shapes are quite different.

The important electrofishing aspects of spherical electrodes (current and voltage distribution in the water surrounding the electrodes) are the lines joining the two electrodes are "current lines" drawn in such a way that the total current between adjacent lines is the same. The lines are very close together near the electrodes, indicating high current density. The dotted lines surrounding the electrodes are lines of constant voltage, illustrating how the total voltage applied to the two electrodes is distributed in the water. The most of the applied voltage is used up in the space immediately surrounding the electrode itself with succeedingly small portions used in regions further removed from the electrode surface. In water at large distances from the electrodes, the current density is very low, there is little voltage gradient and therefore fish will be relatively unaffected. For these reasons, electrode spacing is a relatively unimportant parameter for spherical electrodes. Each electrode essentially acts independently and is surrounded by its own spherical shell or zone of effective current density.

With regard to importance of electrode size, two electrodes, one twice as large as the other was compared. The total voltage applied to the two electrode system is the same (200 V across two identical electrodes). The larger electrode produces twice as large a current density at every point external to the electrode and thus has potentially a much greater effective range. If there is a fish in the field, at the same distance the larger electrode produces twice the voltage.

This desirable effect of increasing the current density and voltage at points external to the electrode is gained at the expense of increased electrode current caused by a reduction in electrode resistance. The same effect could be caused by increasing the voltage applied to the electrodes. However, the larger electrode has a second major advantage by comparing the maximum current densities at the electrode surface. The larger electrode actually has only one half as large a current density at the electrode surface; though surface area of the electrode is four times larger, the current is only doubled. The larger electrode thus offers two major improvements; a larger effective zone (greater range) and a reduced danger zone (lower maximum current density). To achieve these benefits requires an increase in current and power, but operation at the same total voltage. In contrast, increasing the voltage without increasing the electrode size increases both the danger zone and the effective zone and requires a much greater increase in power since both voltage and current are increased. The advantages of increasing the size of the electrodes are such that it is always advantageous to use the largest electrodes possible within the limitations imposed by physical constraints and the electrical limits imposed by the generator and electrical control system.

Although similar in general characteristics, the behavior of cylindrical electrodes is different in degree and in terms of the relative importance of the various dimensional parameters needed to describe their geometry. Unlike spherical electrodes, the spacing between cylindrical electrodes does have important effects even when the spacing is large compared to the radius of the electrode. The important parameter of a long thin cylindrical electrode is the ratio of electrode separation (D) to the electrode radius (R). The resistance between a pair of these electrodes is proportional to the natural logarithm of this ratio. Because this is a relatively slowly varying function, the performance of cylindrical electrodes is only slightly affected by rather large changes in D or R. An extremely important result is that doubling R or halving D has the same net result (*i.e.* reducing separation from 3 m to 1.5 m) will give the same result as changing from a 25 mm to 50 mm diameter electrode. Because of this rather small variation in electrode performance with respect to electrode dimensions, the benefits gained by increasing the size of cylindrical electrodes are not as pronounced as for spherical electrodes. Thus while it is still true that a larger electrode radius produces a greater current density at external points and a reduced maximum density at the electrode surface, the improvement is much smaller than for corresponding changes in spherical electrodes. The table below illustrates the magnitude of these changes for purposes of comparison.

Effect of Changes in Electrode Size and Spacing at Constant Voltage

Nature of Change	Present Change in Maximum Current Density at Electrode		Present Change in Current Density at External Point		Present Change in Distance Out to Constant Current Density	
	*Sphe.**	*Cylin***	*Sphe.*	*Cylin.*	*Sphe.*	*Cylin.*
Double radius	-50%	-42%	+100%	+16%	+41%	+16%
Triple radius	-67%	-57%	+200%	+28%	+73%	+28%
Quadruple radius	-75%	-65%	+300%	+38%	+100%	+38%
½ Seperation	No change	+16%	No change	+16%	No change	+16%
1/3 Seperation	No change	+28%	No change	+28%	No change	+28%
¼ Seperation	No change	+38%	No change	+38%	No change	+38%

* Results for spherical electrodes assume separation many times larger than radius. ** Results for cylindrical electrodes assume initial ratio of D/R equal to 150, *i.e.* 3 meter spacing with 4 cm diameter.

All of these results for cylindrical electrodes assume the length of the electrode is much greater than its radius. The effect of increasing the length of such an electrode is to extend the region which is energized without changing the distribution in radial direction. The total current into the electrode will also increase in proportion to its length (if the voltage is held constant).

If the length of a cylindrical electrode is reduced until it is nearly equal to the diameter, the resultant electrode would closely approximate a spherical electrode. Between this extreme and a long cylindrical electrode is a continuum of intermediate configurations which are difficult to describe in simple terms but which can be bounded in behavior by the two basic configurations.

For a comparison of spherical and cylindrical electrodes in terms of current density distribution, a pair of cylindrical electrodes with a pair of spherical electrodes were chosen so that each pair has the same resistance and hence each has the same total current if supplied with the same voltage. The normalized current density for each electrode is shown as a function of horizontal distance from the center of the electrode. The cylindrical electrode has a slightly higher maximum current density (8 per cent larger) but that the spherical electrode has a larger current density for all points out to a distance of about 1.2 m. Beyond this distance the cylindrical electrode current densities is also higher.

The relative advantages of a spherical electrode over a cylindrical electrode are; the spherical electrode produces lower current densities in the "danger zone", higher current densities in the mid-region or "effective zone" and lower current densities in the far region or "perception zone". These effects are a result of the more uniform spreading of current around a spherical electrode and lead to generally superior performance for spherical electrodes. But, spherical electrodes have many disadvantages associated with providing the desirable mechanical features of easy assembly and ability to negotiate obstructions in water. For these reasons, some of the most successful practical electrode systems utilize arrays of cylindrical electrodes interconnected to approximate spherical electrodes. Such arrays obtain the desirable electrical properties of spherical shapes while maintaining the advantageous mechanical properties of cylindrical electrodes.

From the basic phenomena described in the previous sections, a basis for the rational design of effective system, emphasizing the critical parameters in electrode design include;

1. Each electrode is surrounded by a region of steadily decreasing current density which normally includes a "danger zone" close to the electrode, an "effective zone" at moderate distances and a "perception zone" at large distances.

2. Large electrodes (cross section dimension) have three desirable properties; (a) lower resistance hence more current at a given voltage, (b) a larger "effective zone" because of the increased current density at moderate distances, (c) a smaller "danger zone" because of increased surface area and resultant lower maximum current density.

3. Electrode separation is a relatively unimportant parameter whenever the separation is large compared to the cross section dimension of the electrodes. The relative importance of electrode dimensions depends upon the specific electrode. Spherical and thin cylindrical electrodes

offer boundary cases which are useful in estimating actual electrode performance. An electrode in which all dimensions are of the same order of magnitude will behave similar to a spherical electrode (rings, square plates) whereas electrodes having dimensions which are orders of magnitude apart will behave similar to cylindrical electrode (long thin cylinders, long thin plates). For spherical electrodes; (i) the resistance is dependent on electrode radius and varies in inverse proportion to this dimension. (ii)For separations large compared to the radius, each electrode operates independently of all others. (iii) Increased electrode radius causes a large increase in the "effective zone" and a large decrease in the "danger zone". For cylindrical electrodes; (a) the resistance depends upon the ratio of separation D to radius R and upon the length. The variation is very slight in terms of changes in the ratio of D/R and large with respect to changes in length (in inverse proportion). (b)Increased electrode radius causes a moderate increase in "effective zone" and a moderate decrease in "danger zone". (c) Decreased separation causes a moderate in crease in both effective and danger zones.

4. Increased length simply extends the field farther into the water. Spherical electrodes have generally superior electrical properties, but have many mechanical disadvantages. Thus electrode arrays which achieve the benefits of spherical electrodes while utilizing cylindrical elements are often the most effective.

Water Conductivity Effect

There are two major effects associated with changes in water conductivity; the direct effect upon the total current taken by the electrode system, and a less direct effect upon the division of current between the current distributed in the water and fish in the region affected by the electrodes. The most significant problem occurs as the conductivity decreases with a corresponding decrease in electrode current. At some point the current densities in the water become so low that the "effective zone" is reduced to unacceptably small proportions. The only solution is to raise the current densities by raising the applied voltage or by increasing the size of the electrodes.

Fortunately the two effects mentioned above are at least partially counteracting. As conductivity decreases a fish becomes a relatively better conductor and tends to concentrate a larger portion of available current in its body. For this reason it is not necessary to keep the current constant to maintain effectiveness as the conductivity decreases. This is extremely important since to maintain constant current over a large range of conductivity would require very large changes in voltage and electrode size. Earlier work reported by Cuinat (1967) and by the Novotny and Priegel (1971) on stream shocking with DC indicate that for a ten-fold decrease in conductivity the total electrode current can be allowed to drop to one-third of the original value without serious loss of effectiveness.

There is another conductivity related factor which should be noted. When the conductivity is low (below 100 micromhos/cm), fish is generally much more conductive than the water. In this situation when the fish actually touches or comes very close to the electrode, the fish's body becomes an effective extension of the electrode and is therefore, exposed to high values of current. This creates a condition in which harmful levels of current are likely to occur with possible lethal results for fish. In effect, the "danger zone" is expanded, particularly for large fish. This undesirable situation may be partially rectified by using large electrodes to minimize the "danger zone". A large electrode also minimizes the effect of touching the electrode because as its size is increased, the body of the fish becomes a relatively smaller electrode extension with relatively smaller concentration of current in it.

Extremely high conductivities are also troublesome as it becomes necessary to supply large currents to the electrodes. With the equipment used during the study, conductivities over 500 micromhos/cm caused problems of overloading the main generator unless the electrode size was reduced to quite small. One potential solution to this problem lies in utilizing step-down transformers to permit supplying very large currents at reduced voltage.

Electrode Design in Alternating Current (AC)

While it is true that large electrodes producing somewhat spherical current distributions are desirable in any electrofishing system, there are some important differences between AC and DC electrode systems. These differences are due to their producing different reactions in fish.

In an AC system the primary reaction is electrotetanus or simple immobilization of fish. Since the current is of alternating polarity, all electrodes of a particular system are equally effective and thus an AC electrode system should be designed to utilize all electrodes. In a three-phase system this provides one "effective zone" around each of the electrodes in the array. Often three electrodes are essentially identical, but this is not necessary as long as the total current into each electrode is approximately the same. The need for nearly equal currents is dictated by the three-phase generator, which is designed to supply a balanced (equal current) load. Moderate unbalance (up to 30 per cent) is acceptable but the full capability of the generator is not utilized under these conditions. In some instances generator problems can occur due to additional losses in the rotor windings as a direct result of severe unbalance, In general, the electrode system should be matched to the power source to operate about 80 per cent of full rated current. Voltage control with transformers or electrode size adjustments to accommodate varying water conductivity should be provided.

The relative disadvantages of cylindrical electrodes are not as important with AC as with DC. This is again a result of the simple immobilizing reaction produced with AC. Thus, the essentially cylindrical current distribution of cylindrical electrodes producing an effective zone along the entire length of the electrode can be used even if the electrode reaches down to considerable depth into the water. As long as the immobilized fish are accessible to the net handlers, the downward penetration of the current distribution is useful. With DC, where the primary reaction is to attract fish to the surface, this downward penetration of an essentially uniform field is a disadvantage since fish attracted to the lower end of the electrode are not easily captured in weedy or turbid water. In general, however, arrays of cylindrical electrodes grouped to produce the effect of a larger, more spherical, electrode are more effective than single cylindrical electrodes. The specific arrays found to be reasonably effective during the study.

Electrode Design in Direct Current (DC)

In a DC electrofishing system the desired fish reaction is the more complex electrotaxis or attraction to the anode. Since only the anode is effective, a DC electrode system should always be designed with the largest possible cathode and the anode sized to match the generator output. This arrangements minimizes the electric power associated with the cathode, which is wasted since it does not contribute to the desired electrofishing effect. Since the cathode current distribution is unimportant any convenient shape of electrode can be used. What is important is size and spacing between the separate elements if the cathode consists of a number of separate but electrically connected electrodes. Since individual elements which are closely spaced tend to interfere in the sense that they must utilize the same space to distribute the cathode current, much lower cathode resistance is obtained by keeping individual elements well spaced. This is the same phenomenon that occurs in large flat electrodes in which much of the central area can be omitted (leaving only the material along the perimeter) without greatly increasing the resistance or reducing the total current. Arrays of well-spaced cylindrical electrodes fulfill requirements very well.

One purpose of a DC electrofishing system is attracting fish to the surface when water clarity or vegetation growth makes capture of fish at any significant depth difficult. As a result anode must be carefully chosen to produce maximum current densities close to the surface. Essentially this makes spherical electrodes mandatory (or requires cylindrical electrodes to be mounted horizontally). Since a ring-shaped electrode is close to a spherical shape and can be placed essentially at the water surface, it is an effective anode shape. To make the electrode reasonably able to negotiate obstructions, a ring-shaped array of short cylindrical electrodes can be used.

Instructions for Field Study

1. A conductivity meter to measure the water conductivity at the time of electrofishing operations for a given body of water is essential. Knowledge of water conductivity will allow the operator to adjust the voltage source and change electrode size to conform with the conductivity to ensure maximum current densities during operation.

2. Metering devices are essential so that the operator knows that sufficient currents are being provided.

3. Operation at highest allowable current produces maximum effectiveness whenever possible. Adjust for about 80 per cent of this value to allow for normal conductivity variations.

4. Large electrodes and low voltage operation is superior to operation at higher voltages. Use higher voltages only when low water conductivity makes it necessary.

5. The safety switching system can be used as a control to turn the power on and off. This can be effective in approaching areas of possible high fish concentration without power on and then energizing at appropriate time. The net handlers can exercise this control.

6. Knowledge of the lake topography and the habits and habitat of the fish species sought along with some planning before the initial field work can improve the success of electrofishing.

7. It is generally more feasible to fish at night, especially for those species that move towards the shallows at night.

8. Make sure that all electrical systems (power, safety and auxiliary lights) are in good working condition, especially the safety system, before operations begin.

AC Operation

1. Fish are not attracted by AC, but this type of current has the largest range of action, especially in clear, shallow water with a sand or gravel substrate.

2. Occasionally, better performance is obtained at reduced output if fish are being stunned too far from the electrodes.

3. It is necessary at times to operate at high boat speeds especially when seeking northern pike and muskellunge.

DC Operation

1. Fish will be attracted to the anodes without stunning. This can be advantageous when water clarity is poor, vegetative growth is dense, or algae prevents reasonable visibility.

2. The range of DC is minimal. Low pulse-rate pulsed DC is usually better than continuous DC.

3. It is desirable to operate at very low boat speeds to allow the fish to swim up to the anode and remain there until netted. When excessive boat speeds are utilized fish will be attracted to the anode but won't hold for any extended period as they quickly become exhausted, sinking or drifting back under the boat.

PDC Operation

1. Low pulse rates (5-15 pps) are similar to DC except they have greater effective range.

2. Higher pulse rates (40-80-120) have greater range but fish will be stunned as they approach the anodes.

3. Pulsed DC offers the same advantages as DC when operating in turbid water or dense vegetation.

4. As with DC, slow boat speeds are mandatory to fish effectively.

5. Pulse duty cycle appears to have only secondary effects. Settings at 25 and 50 per cent appear to give similar results and hence 25 per cent is preferable since it conserves power. The 10 per cent setting appears to be less effective.

6. Some species selectivity is possible with pulsed DC. Although only qualitative observations have been made, the following may be useful. Trout, carp, bullheads and largemouth and smallmouth bass respond well to higher pulse rates and will approach quite close to the anodes before being stunned. Walleye, yellow perch, bluegills and white and yellow bass are more easily stunned and lower pulse rates are required to bring them close to the anodes. There appears to be an optimal pulse rate for each species; however, quantitative data are lacking.

Operational Methods

The operational systems developed and tested during the study are as follows:

Electrofishing Boat

To extend the usefulness of the pulsed DC boat, the design was modified such that AC electrofishing was also possible under control of a simple mode selection switch on the operator's controls. The resulting combination AC-pulsed DC boat has proved to be a versatile highly effective electrofishing system capable of adjusting to a wide range of conditions.

Mechanical Configuration

A 5.5 m (18-foot) boat was used for electrofishing in the study.

The safety railing surrounding the front deck was provided to facilitate the collecting operation and can accommodate two operators. This railing system also serves as a support for the main lighting system, bow running lights and the safety on-off controls. The railing was extended along each side of the boat all the way to the stern. Entry-ways just behind the front deck permit easy loading but are closed off by a sliding rail during the operation. The full railing affords protection against the hazard of falling overboard into the electric field near the boat.

Floor mat switches were permanently installed on the front deck and require each operator to be in position before the system can be energized. Non-skid surfaces on these mats are essential. The seat mat switch at the rear operator's position was also permanently installed on the rear seat surface.

The booms were adjustable for height and spacing by means of pin-locked adjustments. During transit the booms are slide back into the boat and held in place at the bow by the boom support collars. When metal booms are used, an electrical ground wire terminated with a battery clamp is provided to assure a positive electrical ground for each boom.

All electrical circuits are enclosed in metal conduit with separate conduit systems for the main power (high voltage) circuits and the auxiliary power and safety (low voltage) circuits. The conduit systems are electrically tied to the boat hull by the mounting hardware as are all other conducting objects any where in the boat. Water tight junction boxes were used throughout the electrical system. All points where connecting wires leave the conduit system are equipped with cord connectors to provide strain relief and a weather seal. The plugs and connectors for the electrical system were marine grade with voltage and current ratings in excess of the actual system design values. The manual safety switches were momentary contact push button switches weather sealed by rubber covers which transmitted the actuating force.

The main operating lights were sealed beam lamps with individual switches. Auxiliary power was provided at several outlets, providing considerable flexibility in the location of auxiliary equipment and components. Fire extinguishers, life jackets, hand tools, fuses and other small parts for emergency repairs were part of the normal equipment.

Because the nets used to capture fish must be dipped into the water near the electrodes where the electric fields are large, it is extremely important that the net handles be constructed of materials with good electrical insulating properties. Conducting materials (metals) or organic materials which absorb water and become conductors must be avoided. Although fiberglass covered metal handles have been used, they can cause accidents if the fiberglass covering is damaged, allowing contact between the operator and the metal handle. This completes a circuit from the water through the operator to the boat and can result in severe electrical shock.

Electrical System

The engine-generator unit provides main power and battery charging power. In addition, a number of additional electrical components were mounted in a control box on the generator to provide a self-contained, dual voltage, electrofishing power supply which can be controlled from buttons on the control box or from external foot or hand actuated low voltage switches. This unit can be used alone for ordinary AC electrofishing or combined with other units to provide greater flexibility.

The components which comprised the unit and the ratings of the components are as follows:

1. Engine- 12 hp with electric starter.
2. Main generator – 4.5 kW, 230 V, 180 Hz three phase self-started generator.
3. Auxiliary generator – 55 A, 12 V automotive type alternator and voltage regulator.
4. Controls – a) Engine throttle to provide speed (and hence voltage) control; b)Engine choke, c)Engine on-off switch, d)Safety circuit on-off buttons controlling a three-pole relay with latching contact (12 V coil), e) Main power disconnect switch (circuit breaker).
5. Metering – a) Main generator voltage 0-300 V, b) Main generator current 0-15 A, c) Battery charging current 60-0-60 A.

6. Protective devices – a) 15 A three phase AC circuit breaker, b) 60 A fuse for 12 V power circuits, c) 2 A fuse for 12 V control circuits.

7. Connectors – a) Three-pole connector for external safety switches, b) Four-pole connector for main generator power, c) Two single pole battery connectors, d) Two single-pole connectors for 12 V power.

The entire unit weighs 113 kg including the fuel tank and muffler. The battery was a standard 12 V automobile battery fitted with a separate battery case and located just forward of the engine-generator unit.

The 12 V system and its operation are identical to that in a automobile. Only occasional monitoring of battery water level and observation of the battery ammeter to verify proper operation is necessary. Depending upon the size and quality of the battery, auxiliary equipment can be operated for reasonably extended periods without operating the main generator. When the main generator is operating, approximately 600 W are available at 12 V.

Safety switching is provided by the three-pole AC relay connected to provide self-latching through a holding circuit. The mat switches and safety off buttons are connected in series in the holding circuit such that if any switch is momentarily opened the relay opens disconnecting the main power. The on buttons, one on the generator control box and one remotely located at the bow, are connected to complete the latching circuit. Thus, to initiate operation each operator must be at his assigned station (closing all mat switches) and the operator must press one of the on buttons. The circuit remains energized until any one of the safety mats or off buttons is momentarily opened.

Transformer Unit

To increase the flexibility of the overall system the transformer has been designed as a separate unit which can be used with or without the pulser. For DC or pulsed DC operation the transformer supplies a variable voltage to the pulser and for AC operation the transformer output is connected directly to the electrode system either through the pulser under control of the mode switch or by simply omitting the pulser entirely.

Voltage settings were selected by operating one of a set of four selector switches. If two or more of the selectors are set to the on position, the unit automatically operates on the lower voltage setting. A single voltmeter is provided to indicate the actual output (secondary) voltage. Three ammeters are connected in the primary of the transformer to serve as a means of monitoring the load conditions of the system. With this arrangement the operator need only be concerned with a single current limit of 11.7 A no matter which voltage setting is employed. The actual output current (if desired) is obtained by multiplying the measured current by the appropriate current ratio from the table below.

Transformer Ratios and Ratings

Setting	Unit	Low	Medium	High
Output voltage for 230 V input	230.0	287.0	345.0	460.0
Output current for 11.7 A input	11.7	9.4	7.8	5.85
Voltage ratio	1	1.25	1.50	2.0
Current ratio	1	0.8	0.667	0.5

The design of the transformer was carried out to minimize the weight of the unit. Only two transformer cores were used (in open delta connection) and the windings were designed to be utilized at essentially at full capacity on all voltage settings. An additional low power low voltage winding was also provided to serve as the power supply for the electronic circuits in the pulser unit. These circuits were brought out through a separate connector to supply the pulser. The transformer input was a four pole connector which mates with the generator output connector. A similar four pole connector provides the transformer output which connects to the pulser or directly to the electrode system.

Pulser

The pulser was a solid state switching system which incorporates a three-phase rectifier, a bistable power switch, and a set of timing and control circuits to drive the bistable switch at the desired switching rates A three-pole relay was used to bypass the rectifier and power switch when an AC output is desired. The unit was controlled by two rotary switches; one controlling the duty cycle of the output pulses and the second controlling the mode or pulse rate. Three duty cycles could be selected during pulsed operation with the on-time of the pulse being 10, 25 or 50 per cent of the total pulse time. The mode and pulse rate switch permits selecting DC or AC operation or pulsed operation at 5, 15, 40, 80 or 120 pulses per second. An off position was provided on both controls.

Three meters were provided to monitor the performance of the pulser. A DC ammeter was connected in the output of the power switch to indicate the average value of the output current. This was adequate for all but the lowest pulse rate where the meter can follow the instantaneous current. At this low pulse rate the average of the meter swing or the reading on DC corrected for the duty cycle must be used as a measure of the current. This was not a serious limitation since low pulse rates were seldom needed. Two volt meters were provided; one connected at the output of the power switch to indicate the average output voltage, and a second connected at the rectifier output to indicate the peak voltage. Comparison of the readings of these two meters during pulsed operation provides a measurement of the duty cycle and was a convenient means of determining if the pulser was operating correctly.

The pulser was internally protected against short circuits by a special fast-acting fuse. The interrupting level of this fuse is chosen high enough (25 A) to avoid interruption under the moderate overloads caused by increase in water or bottom material conductivity or increase in electrode exposure. The semiconductor components of the pulser were chosen to accept such overloads without damage.

With an input AC voltage of 230 V, the output of the pulser was 300 V at 15 A DC based on the 4.5 kW rating of the generator. At higher voltages the output current was reduced to remain within the rating of the generator.

Electrodes

Since the boat was intended primarily for operation with pulsed DC the electrode system was designed specifically for this operation mode. For AC operation the electrode configuration was far from optimum in that only two electrodes were effectively used and the AC currents were somewhat unbalanced. However, the configuration has proved adequate and the advantages of having both AC and pulsed DC available have proved to be significant.

Cathode Array

To minimize the voltage and power associated with the cathode a very large array was necessary. To achieve this large size without severely compromising the handling qualities of the boat, the cathode was designed as a set of ten 1.2 m lengths of 2.5 cm diameter flexible conduit. These electrodes were mounted five on each side of the boat, supported by 3 m length of conduit.

Power was brought out to the electrodes in a 1.2 cm aluminum conduit with a length of flexible conduit to provide for folding the entire array for storage and transit. The wires supplying each electrode leave the conduit system through a cord connector for relief of strain and seal against water. The mechanical load was carried by a chain and insulator fastening each electrode to the supporting conduit. To reduce the electric field near the water surface, the top 20.2 cm of each electrode were covered by a length of heat shrinkable insulating tubing.

Anode Array

To achieve the essentially spherical field suitable for DC electrofishing the anode array consisted of two 0.9 m aluminum rings with a large number of "dropper" electrodes suspended from rings. A conventional boom arrangement supported the two rings approximately 0.3 m above the water and between 2.4 and 3.6 m ahead of the bow of the boat. The rings were normally between 0.9 and 2.1 m apart.

Each of the rings served as a separate anode, providing both mechanical support and an electrical connection for the dropper electrodes that actually carried the current into the water. A switch in the boat permitted energizing just one or both the rings. The rings were doubly insulated from the booms, which were grounded to the boat hull for safety. Power was carried out to the rings by a pair of wires in one of the booms with the cross-piece serving as an electrical connector for the ring on the opposite side.

The dropper electrodes were 15.2 cm lengths of stainless steel tubing supported by a copper wire that is clipped to one of the rings. Care must be used in attaching the wire to the stainless steel tube since the junction will corrode rapidly if allowed to come in contact with the water when the power is on. An insulating sleeve was fitted over the tube to provide control over the amount of surface exposed to the water.

For operation with three-phase AC each anode ring was used as a separate electrode and the cathode was used as the third electrode. This resulted in an unbalanced load on the generator, but if the largest current is kept within the generator rating, satisfactory performance was obtained.

Electrical Characteristics

The electrode system was designed to permit adjustments to accommodate varying water conductivities. Three different adjustments were possible; operating with one or both anodes, varying the number of dropper electrodes, and varying the exposure on the dropper electrodes.

Operating with one or both anodes was useful for initially adjusting the electrode system in water of unknown conductivity and also permitted operation of only one anode in very high conductivity water. Because the two anodes were relatively close together the current with both anodes operating was not twice the current with one anode. The actual ratio depended somewhat on the exposure of the dropper electrodes and was in the range of 1.75 (small exposure) to 1.65 (large exposure). One of the original concepts of the design was to permit adjusting to full output on one anode with DC operation and then to use both anodes under pulsed operation where the load on the generator is reduced by the off periods of the pulsed output. This would take advantage of the power reduction with pulsed operation to increase the current during the on period and thus increase the effective zone of the system. Unfortunately, the capabilities of the power switch were such that operation of these high current levels was marginal and this operating concept was abandoned on these units.

Varying the number of dropper electrodes was not intended as a primary adjustment procedure since it produces significant effects only if the number of dropper electrodes is changed radically. In general, as the number of dropper electrodes is reduced the current density around each remaining one is increased, resulting in excessive power loss in some regions and a possible hazard to fish. For this reason the number was always kept above ten (on each anode) and was most often twelve or more. The only condition under which dropper electrodes were removed is in very high conductivity water (above 500 micromhos/cm) where it was the only alternative to not operating at all.

The primary adjustment mechanism was varying the exposure of the dropper electrodes by moving the sliding sleeve of insulating material. A range of electrode resistance of approximately three to one was available with the 15.2 cm dropper electrodes employed. Even with full 15.2 cm exposure the cathode resistance was only about 12 per cent of the total resistance. Thus at worst, about 88 per cent of the system power was delivered to the anodes where it was useful in DC electrofishing.

The adjustment range provided by varying electrode exposure permitted operation at nominal output (300 V, 12 A DC) over a conductivity range from approximately 150 to 400 micromhos/cm. Below 150 micromhos/cm the voltage must be increased and nominal output can be extended to approximately 70 micromhos/cm at 450 V DC. Below this conductivity the system as designed does not deliver nominal power output because of limitations on voltage. Larger anode rings or longer dropper electrodes could be employed to extend this range. Above 400 micromhos/cm it was necessary to remove dropper electrodes or operate with a single anode.

For AC operation the conductivity range for nominal output was somewhat larger because of the availability of higher voltages (normally limited by pulser components) and because the requirements on electrode configuration were less demanding. Thus the low conductivity limit was extended by the higher voltage available and for high conductivity it was feasible to remove dropper electrodes without causing unreasonable current density distribution. Utilization of a step down transformer would also be of value in extending the high conductivity limit. The degree of current imbalance caused by the unsymmetrical electrode configuration varies with anode exposure from having one current about 50 per cent larger at small exposure to about 25 per cent at full exposure.

Results of Electrofishing Experiments

Fish responded to low pulse rate pulsed DC in the same way they responded to continuous DC except that in the former the range of action was greater. Good response to low to moderate pulse rates (10-40 pps) has been obtained with fish coming to the anodes from depths of 3 meters or more under favorable conditions. This type of response was very useful in turbid water or in areas of excessive vegetation since fish will come all the way to the anode where they were easily visible and easily captured. Other situations in which this mode was valuable include electrofishing in rivers where the water current tends to sweep stunned fish away from the boat or in lake in which rocks or other cover made it impossible to see fish that were stunned in their places of refuge.

As the pulse rate increased fish were affected at even greater distances but tend to become immobilized (stunned) as they drew near to the anodes. The distance at which 100 pps pulsed DC affects fish has been estimated at approximately twice the corresponding distance for DC. Some species (large mouth bass, northern pike, trout, bullheads) seem to have greater tolerance and responded readily to high pulse rates, coming very close to the anodes without being stunned. Other species (walleye, bluegill, yellow bass, white bass) seem more susceptible and were stunned at significant distances (1.2 m or more) unless the pulse rate was kept low.

The greatest effective range (for the same anode current) was obtained with AC. Since the fish were simply stunned wherever they were located when they came under the influence of the effective zone, many fish were too deep to be seen or too far away to be captured readily. However, because of the larger range, AC were useful in many situations and the inclusion of an option switch to AC without any change in electrode configuration was very useful.

When DC or pulsed DC was selected during electric fishing, greater results were obtained when the boat operator moved the boat along at a reduced speed. At reduced speeds the fish, especially smaller ones, will hold at anodes and near the surface for longer period, making capture easier. If the boat was moving at too great a speed, the fish will be attracted to the anodes but will not hold for any length of time as they quickly become exhausted and sink back as the boat is propelled over them. Fishing at these low speeds for northern pike and muskellunge is not recommended as these species apparently detect the electric field at great distances and quickly retreat from the shallows to the deeper water. Increased speeds were necessary to intercept these two species as they darted from the shallows. In shallow waters, AC was usually more effective in capturing northern pike and muskellunge as the effective zone was larger (especially if the substrate is sand) and chances of immobilizing them were increased. A northern pike dashing rapidly from shallow water to DC anodes can easily pass right through the electrical zone.

Smaller fish were usually more difficult to capture with AC because they are harder to see after being paralyzed. With DC and pulsed DC greater numbers of smaller fish can be captured since they were not as easily paralyzed and as they swam towards the anodes, they swam out of the protected areas in vegetation and were readily seen. The boat must be moving at low speeds to realize these benefits.

Some fish species not often seen in AC boom shocking were readily taken with pulsed DC (bullheads especially in deep water, more than normal numbers of largemouth bass, lawyers). The pulse rate tends to allow some selectivity in fish capture in that some species were so easily stunned at high pulse rates, walleyes in particular, that capture was unlikely in deep water. Some species were so tolerant of electric effects that they could be brought up from significant depths by high pulse rates (bullheads) without being stunned until they drew very near the anodes.Grass pickerel attracted out of aquatic vegetation submerged 1.8 m and brought to the surface where capture was easy. In contrast, lake sturgeon were readily attracted from deep water to the anodes but were such powerful swimmers that they passed through the electric field making capture impossible. Operating with AC, lake sturgeon were readily taken because they were stunned some distance from the boat.

Unsuccessful Experiments

A number of experiments were unsuccessful or impractical and were discarded as operational concepts. The most significant of these are mentioned below.

Several modified forms for the anode array were evaluated, including a linear array mounted below each boom, an additional cylindrical electrode suspended at the middle of each anode ring, and two anodes operating essentially adjacent to each other. The linear array was definitely inferior in performance despite its increased current capability, apparently caused by the essentially cylindrical current distribution, which was poorer in DC case where attraction to the anode was desired. A long cylindrical electrode was added at the center of each anode ring to increase the anode size for operation in very low conductivity water. The method was not particularly useful since a very long electrode was needed to produce a significant change and the fish were often attracted to the lower end of the electrode where capture was difficult. For AC operation, or for operation in very clear water where attraction to the anode is less important, the concept was of some use. The two anode rings were operated very close together to simulate a single large anode and the increase of depth at which fish could be affected was evaluated. The desired result was achieved but the increase in downward range was small and the loss in sideward range was large. The concept was of limited use because of the inconvenience of modifying the anode arrangement.

Since it is obvious that large numbers of fish perceive the low levels of current density well out ahead of the boat and escape before coming into the effective zone, the possibility of automatically switching the power on and off at a very low rate was evaluated. In principle, this would allow the boat to approach fish without warning. The concept was found impractical since the off period has to be long and fish easily escape during this time, especially with DC or pulsed DC. It was much more effective to switch the power on and off manually.

An attempt was made to combine the desirable attracting capability of low pulse rates with the immobilizing capability (near the anodes) of high pulse rates by providing a switch for the net handler, permitting the varying pulse rates in response to differing conditions. Thus if a large number of active fish were attracted to the anode, a short period at a high pulse rate could be used to immobilize the fish and facilitate capture. The concept has some value and the boats were equipped to operate in this mode. The major difficulties lied in the rapidity with which events took place and in learning to use the technique.

The other concepts like, operating with various on-off switching rates of the main pulse rate, switching off half of the cathode array to obtain better balance of currents during AC operation, and several minor variations in electrode configuration were evaluated and found to have little value.

Guidelines for Operation

In any type of electrofishing equipment, the skill and knowledge of the operators are major factors determining overall effectiveness. In the course of evaluating the performance of DC and pulsed DC operation, a number of differences in operating technique as compared to AC electrofishing were found to be necessary. The most significant difference was the necessity of moving the boat much more slowly than was common in AC electrofishing to allow time for fish to respond. This was particularly important when fish were attracted from deeper water.

Since the operator had several options available (changes in pulse rate or operation with AC) it was necessary to become familiar with the characteristics of each operating mode and select the proper one for the particular operation. In general, AC operation was preferable in shallow clear water where visibility is no problem and it is not necessary to attract fish out of their refuse areas. These situations occur most often at night when fish were in shallow water and foraging and were hence not in refuge areas. These situations occurred most often at night when fish were in shallow water and foraging and were hence not in refuge areas. Pulsed DC operation was most useful in deep or turbid water or where fish must be attracted out of refuges to make them visible. In general, low pulse rates must be used where attraction all the way to the anode was required. Higher rates gave greater range but result in fish being stunned at some distance from the anode.

Some species selectivity was possible because of this phenomenon, since at high pulse rates only certain species will approach close to the anodes. Trout, largemouth bass, carp and bullheads responded well to higher pulse rates and approached quite close to the anodes before being stunned. Walleyes, white bass, yellow perch and bluegills were more easily stunned and lower pulse rates were required to bring them close to the anodes.

There appeared to be an optimal pulse rate for each species. In general, the following pulse ranges have been found useful; trout, largemouth bass and carp (40-120); bullheads (80-120); walleyes (5-40) and white bass, yellow perch and bluegills (15-40).

In certain situations special techniques were proved useful in increasing effectiveness. Rivers should be electro-fished by moving as slowly as possible downstream. Drop-offs and "holes" should be worked over slowly by circling the area several times. Dense vegetation can be sampled by placing only the anodes over the vegetation to attract fish to the surface. Often approaching with the power off and energizing only after attaining the desired position was very effective. Operating with one anode off to concentrate fish under anode when working with only one net handler was sometimes effective.

Summary of the Field Tests with the Electrofishing Experimental Boat

The field tests with the experimental boat have demonstrated that pulsed DC electrofishing holds promise as a significantly improved technique for electrically sampling fish populations in lakes and rivers. In particular, the test had demonstrated the following specific points regarding pulsed DC electrofishing.

1. Comparing the distance at which fish were affected (size of effective zone) for equal peak anode current, pulsed DC was intermediate between DC and AC. The effective zone for high pulse rate (100 pps) was estimated to be twice as large as for DC.

2. Pulsed DC was a solution to the problem of electrofishing in turbid water or in areas where excessive vegetation made AC electrofishing ineffective. Attracting fish to the surface near the anode required a properly designed anode and operation at suitable (low) pulse rates.

3. The depth to which effective electrofishing was possible could be extended by using pulsed DC. In general, higher pulse rates (100 pps) were needed since the effective zone was larger with high pulse rates.

4. If it is necessary to attract fish all the way to the anode, the pulse rate must be low (less than 40 pps with the simple rectangular pulses) to avoid stunning fish as they approach anode.

5. Some species selectivity was possible with pulsed DC. High pulse rates stun certain species readily, producing low numbers of these species close to the anode and relatively larger numbers at some distance from the anode.

6. With the experimental equipment, the option to operate with AC was a significant advantage since the size of the effective zone was largest with AC.

7. Once familiar with the characteristics of pulsed DC and AC electrofishing, operators can take advantage of the flexibility of having both systems available to meet the varied demands of sampling in lakes and rivers.

Electrofishing Boat using AC Current

Many of the concepts and electrical design changes developed in connection with pulsed DC boat were applicable to AC boats. The work on AC electrofishing had been confined primarily to transferring the concepts developed in the work on pulsed DC to AC systems with emphasis on AC electrode design and development of a special transformer for use in AC electrofishing. Guidelines for operating AC electrofishing boats over a wide range of water conductivity were also developed based on actual field trials of prototype boats.

Mechanical Configuration

The AC electrofishing boats utilized for trials was a standard 1.8 m aluminum boat. Considerable variation in component location, boom support structure, and auxiliary equipment existed since boats were modified to suit the individual needs of the local situation.

The safety disconnect system and safety switches (including mat switches), electrical interconnection system, lighting and auxiliary power system and special consideration for the insulating qualities of the net handles were the same as that of pulsed DC boats.

Electrical System

The electrical system of the AC boat was essentially the same as for the pulsed DC boats except that the pulser was omitted and the connections to the electrode system were modified to properly supply the AC electrodes.

Engine-Generator and Generator Control Unit

With the exception of minor modifications in the later units the engine-generator and generator control unit were identical as that of pulsed DC boat. The later units had a key start-switch instead of start-stop buttons for the engine and some units would be equipped with lights for night operations.

Transformer Unit

Since the AC boats were not restrained by the current limits inherent in the pulser, the transformer was redesigned to better match the generator power rating to the conductivity range over which the boat must operate. To cover the widest possible conductivity range the transformer was capable of either increasing the voltage for low conductivities or decreasing the voltages for high conductivities. The table below gives the voltage and current ratings and transformer ratios available on these " raise-lower" transformers.

Transformer Ratios and Ratings for "Raise-Lower" Transformers

Setting	Lower Low	Lower Medium	Lower High	Raise Low	Raise Medium	Raise High
Output voltage For 230 V input	92.0	161.0	230.0	322.0	391.0	460.0
Output current For 11.7 A input	29.2	16.7	11.7	8.35	6.9	5.85
Voltage ratio	0.4	0.7	1.0	1.4	1.7	2.0
Current ratio	2.5	1.43	1.0	0.715	0.59	0.5

Except for the addition of lower ranges and the modification in transformer ratios the transformers were the same as those used on the pulsed DC boats. All transformers had the ammeters connected in the primary circuit so the operator need only be concerned with the single current limit of 11.7 A imposed by the generator rating. The actual output (electrode) current must be obtained by multiplying the measured current by the appropriate current ratio.

Electrode System

Electrofishing with AC was primarily restricted to shallow clear waters. Since fish were normally only found in abundance in such areas of a lake after dark, the best success with AC electrofishing was obtained by night operation with appropriate lighting arrangements to permit visual observation and collection of stunned fish. Normal above water flood lighting illuminated a large area surrounding the electrodes. It was also characteristic of AC electrofishing to operate the boat at the maximum speed permitted by the ability of the net handlers to observe and capture fish stunned by electric current.

The combination of night operation and high boat speed called for an electrode system which maximized visibility of stunned fish and was capable of readily negotiating obstructions in the water. An electrode which produced a significant surface water turbulence was a great disadvantage since it caused much of the incident light to be reflected off the water surface and greatly interferes with visibility into the water. Compared to the requirements of a DC electrode in which the primary need was to keep the maximum current density region near the water surface to attract fish to the surface, an AC electrode was much less critical in terms of current distribution and much more critical in terms of the physical factors of low turbulence and ease of negotiating obstructions (because of higher boat speed). These considerations tend to favor the use of cylindrical electrodes in AC electrofishing.

The major disadvantage of a cylindrical electrode was the relatively poor current distribution and the associated difficulty of making the electrode large enough to avoid excessively high current density regions. Simply lengthening a cylindrical electrode does not alter the current distribution but only extends the same distribution to a larger region. Array of cylindrical electrodes properly arranged to minimize high current density regions and at the same time minimizing water turbulence had been found to be a good solution. Such arrays made possible the increased electrode size necessary for effective electrofishing without the need to use large diameter electrodes.

Cylindrical Electrode Arrays

The arrays developed during the study utilized two basic concepts to produce the desired electrode system characteristics; paired electrodes and in-line arrays. A paired electrode is simply two thin cylindrical electrodes supported by a separator which places the two electrodes 4 to 6 diameter apart. The pair functioned as a unit and can be combined with other similar units to form a more complex array. The mounting to the boom included a swivel to permit the pair to rotate to minimize turbulence. A paired cylindrical electrode produced the effect of a large diameter electrode without the added weight and drag of a larger diameter electrode. A in-line array is simply an arrangement in which all electrodes are mounted directly under the boom in a straight line. This had an obvious advantage in reducing turbulence as well as making mounting a relatively simple matter.

Out of the two array utilized in the study, one was developed by Leon Johnson based on extensive field tests of various arrangements designed to increase the size of the overall electrode system. The multiple pair was a modification of the arrangement designed to expand the lateral dimension of the effective zone where a wider fishing zone was required. It appeared that both configurations were useful depending upon the conditions and species being sought. Each array used the same basic electrode element; a paired electrode consisting of two 1.8 m length of 2.5 cm diameter flexible conduit separated approximately 10 cm apart. Normally six elements were utilized, but in very high conductivity water it was possible to operate with only three elements, removing one from each of the terminals of the generator. Occasionally a single cylindrical electrode was used in place of a pair, but this was much less effective and often produced hazardous regions close to the electrode.

The boom support structure was adjustable for boom height (above water) and boom separation. Normally the booms were adjusted to operate about one foot above the water. This adjustment was necessary to compensate for changes in load and load distribution in the boat. Boom separation adjustments were used partly for convenience in working along shorelines or weed beds and partly to compensate for changes in water conductivity. The separation was variable from a minimum of 1.8 m to a maximum of 4 m (measured at tip of booms). In general wider separations yielded a wider lateral dimension for the effective zone. However, in low conductivity water the width of the effective zone of each electrode could become small enough to create a region near the center of the booms in which fish were relatively unaffected. In these cases the separation must be reduced.

Electrical Characteristics

Each of the arrays produced an essentially balanced electrical load. The alternate polarity array had a lower resistance because the individual elements connected each phase are farther apart and hence did not interact with each other to as great extent as for the multiple array. However, even with the 3-meter separation between elements of the same phase, there was a significant interaction and the current density distribution was much improved over that which was obtained in a three element array.

The interaction effect mentioned above could be observed by comparing the current for the three electrode array and the six electrode alternate polarity array. Without interaction the six electrode array would have twice the current of the three electrode array, while the actual comparison was 6.0 to 3.3 or about 180 per cent. The larger interaction effect of the multiple array was shown by the corresponding comparison of 5.2 to 3.3 or about 158 per cent. These interaction effects were generally desirable since they reflected a reduction in the maximum current density and hence a reduced danger zone close in to each electrode element.

Two other important characteristics of the arrays were that boom separation was clearly a relatively unimportant parameter in terms of total current delivered to the electrodes. This was characteristic of cylindrical electrodes. The importance of boom separation lied not in the total current but in controlling the extent to which the effective zones of the separate electrode elements overlap. Too wide a separation would create non-overlapping effective zones in the center area between the booms. The multiple array had some advantage in this respect since the effective zone around each of the multiple electrodes was larger and hence wider boom separation were possible.

The second characteristic was the effect of boat speed on the total current. Higher speed should result in lower currents since the electrode elements were deflected upward by the drag in the water. If the electrode elements were to ride along the surface only half in the water, the current could be reduced by approximately 50 per cent. Measurements of the actual reduction in current were much less. The reduction was seldom more than 15 per cent and was not a serious problem with the electrode elements used.

Electrofishing Results

Most of the field evaluation carried out on AC boats had been concerned with performance in low conductivity water and had used the alternate polarity electrode system. The data clearly showed the advantage of operation at higher voltage when the conductivity was below 10 micromhos/cm. For conductivities below 40 all evaluations were "ineffective" except those at 390 and 460 V. The lowest conductivity tested was 14 micromhos/cm where reasonable effectiveness was obtained at 460 V and the effectiveness was extremely poor at 230 V.

Since effectiveness was increased with increased current, operation at the largest possible current within the rating of the generator should be sought under all operating conditions. The transformer settings available provided operation from a minimum conductivity of 10-15 micromhos/cm up to about 1200 micromhos/cm with the six electrode array. By removing three electrodes the upper limit could be extended to over 2000 micromhos/cm. The Table indicated the proper choice of voltage level and number of electrodes for various conductivity ranges.

The best approach was to use the six electrode array and operate at the voltage which most nearly utilized the full 11.7 A rating of the generator. Electrodes should be removed only if generator overloading is a problem at the lowest voltage setting of the transformer.

Voltage and Electrode Selection

Conductivity Range (micromhos/cm)	Voltage (V)	Number of Electrodes
10-50	460	6
50-70	390	6
70-100	320	6
100-200	230	6
200-400	160	6
400-1200	90	6
1200-2000	90	3

The multiple pair array had received only limited field use but initial results were very encouraging. The array had a much larger lateral effective zone with initial estimates indicating about 35 per cent increase in the overall width of the effective zone at the same anode current. Since the multiple array had a higher resistance it must be operated at higher voltage to attain the same current as for the alternate pair array. It was therefore well suited to intermediate and higher conductivities where full output current was readily attainable. The multiple pair array produced a lower maximum current density and hence a smaller hazardous zone than the alternate array and should therefore reduce the chances of injuring fish.

The two arrays produced somewhat different current distributions, each offering some advantages. The alternate array tended to produce an elongated and relatively narrow effective zone along the entire length of each boom. Normally the effective zones associated with each boom overlapped in the center region between booms, but if boom separation is too large there was often a zone of insufficient current density in the center region. The multiple pair tended to produce a wide effective zone of relatively short length centered about the forward electrode pairs. Previously the width of the effective zone was considerably larger but was achieved by reducing the length of the effective zone. Under the varying conditions of lake electrofishing each of the field pattern had value. For example, in electrofishing along shorelines for northern pike and muskellunge, fish were often captured as they attempted to swim past the boat toward deep water. The longer effective zone of the alternate polarity array was then a distinct advantage. For most other species the wider effective zone of the multiple pair array provided a larger sampling zone and was hence often advantageous.

Since the two array could readily be accommodated on the same boom structure by providing two additional connection points, the design provided for using either array. Additional field testing would be necessary to determine if both arrays were sufficiently useful to be retained. Further development of the multiple pair array was also possible by utilizing three pairs for each forward electrode. But the present arrays already represented close to the maximum feasible boom loading without severely complicating the boom support structure and ultimately the handling qualities of the boat.

Results of AC Electrofishing

The work carried out on improving the effectiveness and extending the conductivity range of AC electrofishing boats which produced the following results;

1. Transformers can be used to advantage in extending the conductivity range of AC electrofishing boats. Successful operation down to conductivities of 10-15 micromhos/cm at 460 V had been demonstrated. Extending the range to high conductivities appeared feasible but has not been adequately evaluated.

2. The value of large electrodes in increasing effectiveness and reducing injuries to fish had been demonstrated. Paired cylindrical electrodes mounted in line along the booms had been advantageous in minimizing water turbulence (hence maximizing visibility) and producing reasonable current density distributions. Operation with the largest possible electrodes at lowest possible voltage was desirable.

3. Alternate polarity and multiple pair electrodes had demonstrated to be usable configurations. It appeared that alternate polarity electrodes were advantageous in very low conductivity water and where a relatively long narrow effective zone was desirable – for example, when fish were to be captured as they attempted to escape past the boat to deep water. Multiple pair electrodes produced a wider effective zone with smaller hazardous zones and offered potential for minimizing injury to fish and increasing the lateral size of the sampling zone.

Chapter 6

Fishing with Electricity in other Countries

Direct Current Electrofishing in Streams of Canada

During the course of investigation of the inter-specific relationship of eels and young Atlantic salmon, direct current electrical fishing gear was used to collect fishes in 1952. The electrical equipment used was like the Wolf machine and its accessories. The operator carried the positive electrode, which was connected to the petrol driven generator by a long cable of insulated wire. The negative electrode consisted of a small square of copper screening. It was attached to the generator by a short cable and was grounded on the river bottom. Its electrical output was1000 watt. In open circuit the speed of the motor was adjusted to maintain 750 volts. Under load, while fishing, the potential supplied and current drawn varied from 600 to 700 volts and from 1.0 to 1.7 amperes in different waters.

2958 fishes, belonging to 14 species were taken by electric fishing. The fish that were captured included species such as, eels, sculpin and burbot, that are usually very difficult to capture with seines. The stream bottoms included types with many large boulders and overhanging banks that would have made seining very difficult and inefficient ways of capturing large proportions of fish present. In these tests the type of stream bottom did not greatly affect the success of electrofishing.

Fish exhibited a marked galvanotropic response, such as, eels particularly and salmon fry were collected most readily. The success of electrofishing depended greatly upon the facts that all the experimental sections were shallow (between 0.3 to 0.6 m in depth) and the electrical conductivity of the water was suitable for the apparatus in use, and provided a fairly wide electric field,

Wolf Power Plant

The Wolf power plant was a portable direct current generator with a direct drive from a small, air-cooled petrol engine bolted to the generator housing. The generator was a re-built, shunt-wound unit. Any of five voltages can be produced by introducing various resistances into the generator field circuit. These resistances and the associated switching device were located in a control box mounted on the top of the generator. The maximum potential supplied was about 610 volts; the minimum unloaded voltage was about 400. Voltages down to 220 was obtained by closing the throttle of the engine. The maximum current that could be drawn varied from about 1.0 amp. at 550 volts 5,0amp. at 250 volts. This variation in power output was due to the generator winding and the power and speed characteristics of the petrol engine. The power plant, attached through rubber mounts to a tubular steel carrying cradle, weighs about 98 kg.

The smaller American unit was a compound-wound generator rated at 110 volts DC and 600 watts at 1750 rpm. For fishing, the speed of operation was increased so that the generator delivered about 160 volts.

To operate the apparatus the negative pole of the generator was grounded by connecting it through a long insulated wire to a 0.6 by 1.5 m piece of non-rusting wire mesh laid anywhere on the stream bottom. The positive generator pole was connected to 100 m of two-conductor cable mounted on a reel which was carried on the operator's back. As the operator moved the cable unwinds and lied on the stream bed. At the end of the reel was a revolving, sliding contact which connected the long cable on the reel to a shorter two-conductor cable fastened to the proximal end of the electrode handle.

Three types of positive electrode were used with the Wolf apparatus. These were essentially loops of 3/8-inch metal attached to heavily insulated shafts of the same material. One type was an oblong grid (28 x 20 cm) with a 60 cm insulated shank. This shank could be locked into the end of a tubular insulated handle which contains an open spring switch. The cable leading from the reel, plugs into the upper end of this handle. Unless the switch was closed by properly gripping the insulated handle the field coil circuit of the generator remains open and no current flowed. Another electrode consisted of a simple aluminum loop (30 x 20 cm) with a 195 cm insulated shaft (dismountable in three sections). This electrode was used with no handle switch. A third type of electrode was a ring of copper tubing attached by cross braces to an insulated cable which had on its proximal end a fixture for attachment to the switch handle.

Collection of fish was made by immersing the positive electrode in water and gathering narcotized fish with dip nets and/or seines. For ordinary work either one of the first two types of electrode was used. The long handled aluminum loop was more convenient for reaching under obstructions. The ring and cable electrode had been used for casual collection in deep holes, off bridges *etc*. It was flung out to the desired spot and was drawn slowly in, some fish being immediately narcotized and others following it within reach of the operator.

The effective field for fishing was between positive electrode and stream bottom up to a depth of 0.6 or 0.9 m, and bulged laterally 0.6 to 1.8 m depending on the conductivity of the water. With the electrode deeply immersed the effective field did not extend above the plane of the electrode.

In waters under 1 m deep most fish could be taken out from an area, uninjured, regardless of obstructions or other fish cover. It was a very effective method for collecting eels of all sizes. Use of DC was preferable to AC, especially where it was desirable to collect the maximum number of fish without injury.

Pulse-Modulated DC Electric Fishing Machine to Capture Eels in New Zealand

A portable electric fishing machine using pulse-modulated DC had been developed by A.M.R. Burnet in 1952 to catch eels, all the members of which were hidden in cover during daylight hours and some of which, the smaller ones, remain in cover all the time. The machine has been proved to be a very useful and efficient tool for the study of eel populations.

Design of the Machine – Physiological Basis

As a first step some experimental work was done to establish the physiological constants and thus the basis for the electrical design of a suitable machine. For the preliminary work, electrodes were placed at the ends of a trough 14 inch wide 7 inch deep and 100 inch long. The water depth was adjusted to give a convenient resistance. The electric current was thus contained within an even, rectangular cross section and therefore the voltage gradient was reasonably uniform.

Theoretically a rectilinear pulse is the ideal shape. But in practice, it was very much simpler to use a condenser discharge.

Using a repetition frequency of 15 c/s, the voltage gradient necessary to produce paralysis was determined for various pulse lengths. The galvanic threshold or rheobase is the minimum continuous DC voltage required to produce contraction of the tissue (*i.e.* paralysis). There is some evidence that a constant potential difference between the head and tail of the fish will produce paralysis, irrespective of the length of the fish. Thus longer the fish the lower the potential gradient required.

The chronaxie is the shortest duration of a current of twice the rheobasic voltage that will stimulate the tissue. The shorter chronaxie time for a fish will indicate a higher speed of action of the tissue, could probably be correlated with the different methods of swimming of the species.

As regard the repetition frequency on the voltage required to paralyze the fish, Kreutzer and Peglow claimed that if the frequency of the pulse is matched to the natural wriggling frequency of the fish, it will be caused to swim up to the positive electrode.

The pulse technique so far investigated has been based upon the summation of contractions. If the time interval between stimuli is reduced to about twice the chronaxie time, then subliminal stimuli will be summated to produce an effect. An investigation of the voltage and pulse times required to produce paralysis at the summation frequency of about 200 pulses per second may lead to useful results.

The results obtained by overseas workers suggest that an output of atleast 500 V would be necessary. It has been found that peak pulse voltages between 600 and 750 V were satisfactory in practice.

Good results were obtained with a machine that has a pulse time equivalent to a rectangular pulse of about 1 msec. The smallest eels encountered 8.5 cm long could be paralyzed within about 30 cm of the electrode. However, for use with salmonids a much stronger field was necessary and an equivalent pulse time of at least 2 msec. should be allowed.

Load resistance had a large effect on the range of the machine. It was necessary to make measurements of circuit resistance, using an electrode set-up of the type decided upon, in the various types of streams in which the machine was used. The water in most of the streams fished had a specific resistance of about 5000 ohms per cubic inch, and with the particular electrode system used the output resistance was about 250 ohms. Thus to obtain a pulse time of 2 msec., RC was 0.0054 and the C was then 20 micro-farad (approx.).

The amount of power required to drive the system was ½ NCV square watts, where N = repetition frequency in cycles per second, C = capacity in farads, and V = output voltage. If C = 20 micro-farad, N = 20 cycles per second, and V = 700 V, then power required = 100 W.

With this pulse length the results obtained were approximately equal to those that would be obtained with a continuous DC machine of half the output voltage. This would use 500 W. There was therefore a considerable saving in power. This low power requirement made a small portable machine driven from an accumulator a practical possibility.

The simplest method of producing a pulse is by using a capacitor discharge. Inductance charging of the capacitor was considered desirable in spite of the weight of the inductance, as resistance charging could not be more than 50 per cent efficient.

The machine which was used had a 20 W, 480 V genemotor, and will produce a pulse output of about 40 W, at 630 V for an input of 85 W.

The power was supplied by a 24 V, unit made up of four 6 V., 17 A-hour motor cycle batteries. The capacity of this unit was sufficient for operation for five hours, when the genemotor was drawing 85 W.

When the electrodes were small compared with the spacing between them, most of the resistance was "end effect", *i.e.* close to the electrode, and correspondingly the voltage gradient was high close to the electrodes and low half-way between.

The lines of current flow in the water followed the same pattern as the lines of force between magnetic poles. Between the electrodes the field bulged out to many times the cross section area of the electrodes. There was thus a lower current density per unit area, and by Ohm's law the voltage gradient per foot was correspondingly lower. Resistances measured with 6 inch square electrodes in water of approximately 5000 ohms per inch. Cube resistivity gave the following results:

Six inch spacing – 230 ohms; 4 feet spacing – 340 ohms; 10 feet – 360 ohms; 20 feet – 370 ohms. A single electrode with earth return (a piece of wire netting 2 feet 6 inch by 5 feet placed on the stream bed) gave a resistance of 250 ohms, and this remained fairly constant for spacing up to 100 yards, except when the two were within a few feet.

Observations on the stunning of eels, using a 6 inch square electrode, earth return system, and 600 peak volt suggested that the voltage gradient is of the order;

1. One foot from electrode = 8 V per inch
2. 3 feet from electrode = 4 V per inch
3. 10 feet from electrode = 1 V per inch

For the study of eel populations it was found that an earth return system to be very suitable. The best earth was found to be a piece of galvanized wire netting, 2.5 by 5 feet placed on the bed of the stream. As this electrode was much larger than the positive one, little of the output voltage was dissipated around it. But this system will not work satisfactorily if the ground resistance is high (rock for example) The positive electrode was a 6 inch square of 1/4 inch brass rod, with two cross pieces. This was attached to a 5 feet waxed wood handle. Connection to the machine was via 100 m of 7/0.029 TRS cable with a spring release switch on the handle.

A number of reactions have been observed during fishing operation. Some of the eels were frightened off downstream when in the periphery of the field. Many of the small eels, less than 25 cm, were frightened out quite close to the operator and the electrode to paralyze them. A large number were paralyzed in cover and carried out by the water current. A number of those not carried out often swim out after the electrode had moved. With very deep cover, the yield was much better on the second run. While this technique has been proved very satisfactory for eels, a two electrode system, with its resultant more even field, may be more suitable for salmonids, particularly if worked upstream. The voltage gradient near the electrodes will not be as high as with the single electrode and earth return. But the load resistance will be higher and consequently the pulse duration will be longer. This will to some extent, compensate for the lower voltage gradient.

A single run with the electric machine had taken between 26 and 71 per cent by number of the population, with a mean value of 43 per cent. Thus, on the average, 67 per cent could be removed with 2 runs and 81 per cent in three runs.

Electrofishing Shocker Developed by the Bureau of Commercial Fisheries, USA in 1966

The type IV shocker was the fourth of the series developed by the Bureau of Commercial Fisheries at the Seattle Biological Laboratory in 1966 for collection of fish.

The type IV shocker consisted of a 12 volt/400-volt dynamotor, the output of which was connected in parallel with a 100 microfarad, 600 volt capacitor bank to provide power for the shocker. In operation, this supply was connected on one set of normally open contacts of a mercury-wetted relay, capable of operating up to

100 pulses per second. The other side of the normally open contacts of the relay was connected to the 2-ampere fuse, power output plug, and cable to the anode. The mercury-wetted relay was driven with an OA4G relaxation oscillator which, when fired, energized the coil of the relay. Duration of the output pulse was fixed by the size of the capacitor in the power supply side of the mercury-wetted relay coil. Frequency was determined by the time constant of the selected capacitance and fixed resistance in the trigger anode circuit of the OA4G thyratron. Frequencies of 20 to 75 pulses per second with a duration of 6 milliseconds were easily obtained with this circuit.

A power switch located on the front panel, when in the "on" position, connected the dynamotor to a 12 volt battery. With the local remote switch in the local position, the unit operated continuously. With the switch in the remote position, the dynamotor operated only when the strip-switch on the anode handle was closed. The strip-switch energized the coil of the 12 volt, double pole, single-throw relay. One set of contacts on this relay also controlled the running of the timer motor.

A normally closed momentary switch located on the front panel, was used to open the plate lead of the OA4G to allow for deionization of the tube and to restore normal pulsing operation after a misfire. A misfire could be caused by accidental shorting of the electrodes in the water.

A four-position selector switch connected different values of capacitance into the trigger anode circuit to vary the frequency of the oscillator.

A running-time meter and a small indicator lamp were installed on top of the unit. The lamp was wired directly across the output of the unit and gave visual indication of the output pulse. The running-time meter recorded cumulative time that the unit ran, to assist in determining catch per unit of effort.

A 12 volt battery with a 50-ampere-hour rating was generally sufficient as a power source for 4 to 5 hours of use.

Fish Reactions to Increasing Current Strength

The alternating directional response to the AC was replaced by a unidirectional one that caused the fish to move progressively toward the anode. This directional response was of primary importance in electrofishing as the fish can be attracted out of swift or turbid water or from heavy underwater cover to a predictable point. Although DC produced the desired galvanotropic effect, it still leaves much to be desired because of the refractoriness of the fish.

Extensive, strong, directional swimming did not occur unless the DC was pulsed. The reactions of fish to pulsed DC of increasing strength were similar to those reported for DC, except that before directional swimming commenced, the pulsations set up a series of movements which were probably galvanotropic reflexes of spinal origin. It was thought that locomotion might be a voluntary escape reaction at that time, but that galvanotropic reflexes produced by pulsed DC inhibited the fish from turning away from the anode. The result was violent milling movements; some fish escaped and others succumbed to galvanotaxis.

Haskell (1957) and Vibert (1963) found that direct stimulation of muscle required considerably greater voltage than was needed for neural reaction. Neural stimulation, therefore, was the most useful stimulus in electrofishing because it occurred at lower electric intensities than muscle stimulation but achieved similar results.

Pulsations below 50 per second attracted fish poorly. The most desirable fish reactions were at 50 to 90 pulses per second except in resistive waters. There, frequencies as high as 100 pulses per second were necessary to produce similar reactions.rates from 90 to 140 pulses per second caused very rapid swimming undulations, but above 140 tended to narcotize. Excessively high frequencies had no apparent effect upon fish.

Although pulsations could set up the strong rhythmical flexures required for galvanotaxis, a certain minimal amount of electrical energy must reach the fish. If the time (duration) of the electrical impulses were extended, the fish experienced prolonged exposure at given voltages and hence was subjected to increased power. As the frequency and duration were electronically set and remained stable, their optimum combination was one of the most important consideration of electrofishing.

As the frequency increased from 50 to 90 pulses per second, the maximum duration could decrease from 12 to 6 milliseconds. Greater durations would produce fish "fatigue" and constrained movement which might represent refractoriness in the nerves and muscles. Lesser durations did not provide the needed energy.

The square wave was considered superior for electrofishing because the maximum voltage was produced throughout its duration. Other wave shapes had a lower efficiency.

The voltage potential of the DC shocker should be appreciable to provide a strong stimulative gradient at a distance from the anode; yet practical limits must be set for the safety of the operators and in consideration of the physics of water conduction of electricity. An output of 150 volts approached the desirable minimum in more conductive fresh waters that characteristically have resistivities of 1000 to 15000 ohm per cubic centimeter.

The amperage in electrofishing provided the quantitative effect of a stimulating current. Because the power that can be transmitted through water was a function of resistivity, the optimum amperage varied with the water.

Type IV shocker indicated that a current of 60 milliamperes (mA)was inadequate but that one of 150 to 180 mA was adequate in waters over 5000 ohm per cubic cm. resistivity; a shocker capable of producing 1 to 2 ampere was adequate in all fresh waters.

The type IV shocker had proven highly successful in attracting fish from distances, and they often forcibly collide with the anode through galvanotaxis.

The electric energy in water could be increased by the use of "optimum-sized" electrodes. The maximum surface area of the electrode is governed by the practicality in manipulating it –especially the anode, or fish attracting pole, which should be highly mobile and of a size that could be passed through swift waters without excessive difficulty. The conductive grid of the anode was best when about 40 cm square. This size of grid actually hindered conduction in most waters, but the increased current densities immediately surrounding the anode facilitated electronarcosis which eased problems of capturing fish.

The electrical resistance inherent with small anodes could be overcome with the use of enlarged cathodes in practical stream operations because cathode could be stationary. The optimum size for the cathode was dependent upon the water resistivity. The suggested minimum cathode surface area for waters less than 30000 ohm per cubic cm. was 2.3 m square, but for more resistive waters the cathode size should be increased.

The most satisfactory cathode was an aluminum boat, which can also be used to transport the shocker and fish holding tubs.

Techniques of Fishing with Electricity

In electric fishing one must wade or float, depending on the depth or swiftness of the water. In suitable waters, the operators wade and can probe the anode into likely fish habitat. Wading upstream eliminated effects of turbidity caused by bottom sediment. Furthermore, if collections were for a food=habit study, stunned prey were not swept downstream and consumed by predators. If the turbidity and predation were unimportant, however, collections could be made more efficiently and less strenuously when moving down stream. The fish were normally oriented upstream, or toward the descending electrical field, and the shocked fish initially induced into flight bolt upstream into higher voltage densities, where they were held. Fish that managed to escape were often captured a short distance downstream. The size of the fish captured by wading operations in large streams was usually less than 150 mm., whereas larger fish were taken in deep waters by the floating method.

The floating method of electrofishing was used when the stream was too deep or swift to wade. The anode was clamped rigidly ahead of the boat, extending into the water. One man guided the boat with oars while one or two operators dip fish as the boat drifted with the river.

Considerations for Electrofishing

The effectiveness of the shocker's output energy was sometimes reduced drastically by environmental or biological factors. The resistive effects of water do not alter the pulse frequency or duration but depressed the electric energy input. The power that reached the fish was then modified by the animal's physiological makeup. Some small adjustments in output power could be made to reduce the erratic actions and escape of fish, but this behavior generally had to be accepted.

The resistivity of natural waters depends on the quantity of ionized salts available to carry the electricity. Salts ionize more freely with the rise of water temperature, and this was more pronounced in waters of low salt content.

Practices which reduce the effect of highly resistive waters by delivering more electrical energy to fish include use of high frequencies and durations and use of square waves with peak voltages of 300 to 400 volts; it helps also to maintain large electrodes close together and to take full advantages of the "surprise effect".

Fish flesh has a certain resistivity that decreased with increasing temperature. It was found that electrofishing in highly resistive waters was mildly enhanced due to the more conductive fish tend to distort the electric field by the absorption of electricity. Theoretically, success of electrofishing should increase with rise of temperature, but workers in the field had noted differently. The experiments indicated that maximum shock response was at the temperature preferred by the fish; the work implied that each fish species has a temperature at which it responded most strongly to electrical stimulation.

Output energies commonly used in electrical fishing are capable of killing fish. Death could occur with or without gross physical damage. Mortalities caused by AC electrofishing were higher than those caused by DC or pulsed DC. Harmful effects from pulsed DC were usually a result of excessive exposure or intense electrical field. Electrical stimulation interfered with or stopped respiration in fish for a period of time producing metabolic deficit. The situation became precarious unless the fish were removed quickly from the stimulating currents into water with optimum temperature and dissolved-oxygen concentration.

Chapter 7

Electrical Fishing in India

In India, electrical fishing was first done in 1962 by K. P. Biswas in Odisha to remove bottom and mud dwelling fishes from earthen rearing ponds having a water depth of 3 to 4 feet. He developed an electric seine to remove all the fishes from the ponds, which were 14 in numbers located in Dhauli, near Bhubaneswar, Odisha. The ponds were of different sizes and dimension. The equipments consisted mainly of two components, (1) alternating current (AC) generator driven by 5 hp water cooled diesel engine, and (2) seine net with electrodes. The generator delivered 230 volts at 1500 rpm.

The current from the generator was delivered to the electrodes attached to the seine net through a switch board consisting of voltmeter, ammeter and a field rheostat.

The electrodes consisted of 50 feet flexible bare copper wire of 7 strands, which ran, one along the foot rope of the seine net leaving a gap of 2 feet in both the sides. The other electrode ran along the body of the entire net, 2 feet below the head rope leaving a gap of 2 feet in both the sides like the other electrode along the foot rope.

The seine net was of 60 feet long and 18 feet wide with 5 mm mesh in the webbings. The head rope was provided with plastic floats at 4 feet intervals and in the foot rope lead sinkers were attached 3 feet apart from each other. Both the electrodes remained under water during seining operation, and the distance between the electrodes depended on the depth of water in the pond. The electrical lines of conduction passed vertically across the water column as the seine net moved from one side of the pond to the other.

The electrodes of the net was connected to the output terminals through the switch board and insulated wire. Four persons on the bank (without wading) dragged the net and the current was allowed to flow through the net-electrodes during the entire period of dragging. To make a comparison, the same net was dragged through the pond without electrically charging the net prior to the operation of electric seine. The catch of both the conventional and electric seine was analyzed for numbers, size and weight, species-wise, except in case of minnows, which were weighted and counted in numbers. The entire catch of the conventional net was released in the pond in the living condition after keeping the records of the catch. An hour later, the same seine was dragged with electric field around the electrodes and the current switched off at the time of hauling the seine net. The catch of the electrical seine was then analyzed for numbers, size and weight species-wise. The electrical resistance of the pond water was measured by ohm-meter per cm³.

Irrespective of the varieties the fish catches were found to be more in electrified net than that of conventional net, which was probably due to the oscillotaxis and electronarcosis responses of the fish, when they entered the vertical electrical field created by the electrodes attached to the foot rope and net body, 2 feet below the head rope. The input unloaded voltage of 230 was reduced to 120 to 135 volts with the load of electric seine while dragging. The input loaded voltage of 120 to 125 gave the best result in water resistance of 9000 to 9500 ohms/cm³ and the better results were obtained in water resistance of 9000 to 15000 ohms/cm³. Benthic fishes like, *Labeo calbasu, Channa* sp.*Glassogobius guiris, Mystus* sp., *Rhyncobdella aculeate, Clarias batrachus* and *Cirrhina mrigala* were found to be comparatively more effective in electrical seine than other varieties.

The same author (K. P. Biswas) in 1964 had done electrical seining in larger and deeper water areas ranging from 0.5 to 13.8 acres having water depth of 5.3 to 20 feet. In all he did electrical seining in 24 water areas for 46 times. The resistivity of those waters varied from 6000 to 19000 ohms/cm³. In those trials, continuous AC and interrupted AC of 2 to 4 per second interruptions were used. The seine net used in those experiments was a framed one with 1.5 mm dia. hemp rope, fitted with electrodes of flexible copper wire to charge the net with electric current during operation. The length of the seine net varied from 60 to 150 m depending on the width of the water area. The width of the seine net was 7.5 m with mesh size of 5 cm. in stretched condition. Lead sinkers of 250 g. each were provided in the foot rope at a 3 m. interval. Round polythene floats were provided on head rope of the net at a distance of 2 m. from each other.

The ground electrode ran along the foot rope of the net leaving a gap of 1 m. from each side of it and the top one was fitted in the same manner I m. below the head rope. The third electrode was fitted in between the ground and top ones along the entire length of the net. In 16 trials only two electrodes were used, in which, the ground and top ones were connected with neutral (negative) and the phase of 220 volt AC respectively.

During the trial with three electrode system, the electrodes at the top and the middle were connected with the phase while the neutral line was connected with the ground electrode. Round flexible copper wire of 23/.0076 was used as connecting cable.

Interrupted AC was produced with the help of a mechanical interrupter connected in between the generator and the control panel. It produced frequencies of 2 to 10 per second with different periods of impulse and pause by the combination of different pinions of the gearing system.

Each time before the operation of net with electric charges, it was operated without the application of electric current to compare the efficiency of the electrical fishing method with that of ordinary net. The catch of the control haul was measured for size, number and weight species-wise after which the entire catch was released back in the same water area to keep the population constant. As the current flowed through the electrodes it created an electric field through the water column in the vicinity of the net which moved from one end of the water area to the other as the net was dragged. Fishes were visibly seen to got stunned on coming within the field and were collected in the net behind as it moved. This reduced the percentage of escape to the minimum.

Increased catch both in numbers and weights have been observed in electrical seine over the control one. The effect of electrical seining was due to stunning of fishes as they came to the proximity of the electrical field reducing the percentage of escape to minimum. Continuous AC had a better effect over interrupted AC. In interrupted AC fish possibly got better chances to escape during the period of pause when the electrical field around electrodes disappeared temporarily. Introduction of third electrode could not give the better catch than two electrodes. Electrical seine in general increased the catch of all varieties of fish over control seine except in case of *Barbus sarana* where more fishes were caught in control seine. Species like *L. calbasu, C. reba, G. giuris, H. fossilis, C. carpio, Channa* sp, *N. notopterus, M. cavasius and M. pancailus* could be caught in electrical seine, when control seine failed to catch them, which indicate that benthic fishes were more prone to electric shocks and exhibited narcosis earlier. Fishes of smaller size groups which escaped through control seine were caught in electrical seine irrespective of nature of current, particularly in respect of *C. mrigala, N. notopterus and Channa* sp.

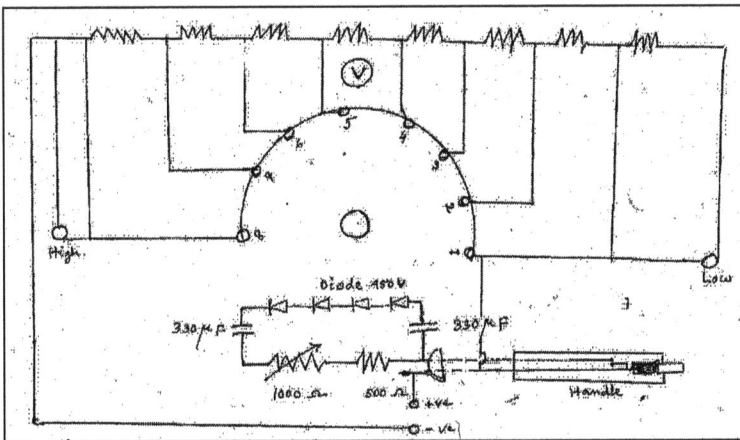

Figure 4: Schematic Diagram of Power Source Use in Electric Fish Catcher in which Switch Breaks the Field Circuit.

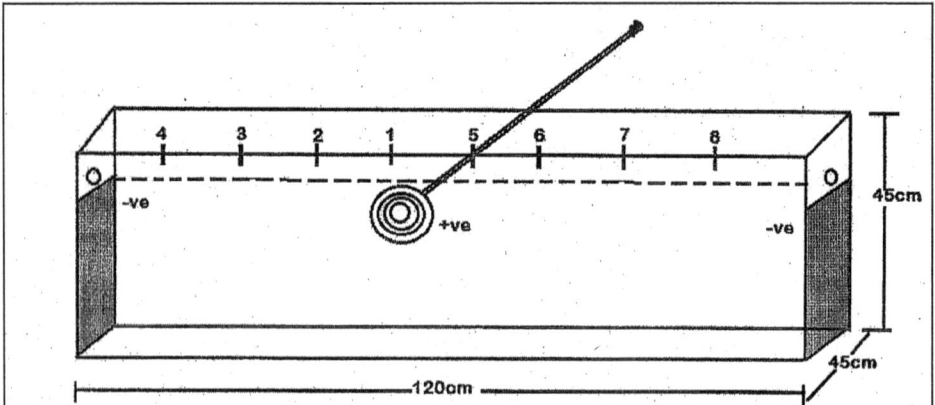

Cather electrode helix coil-half immersed (19cm dia)

Position	Applied DC Voltage (Volts)	Current density in $\mu A/mm^2$
1	55	0.0025
2	45	0.0015
3	40	0.001
4	35	0.0005
5	45	0.0015
6	50	0.001
7	55	0.0005
8	50	0.00025

Figure 5: Field Pattern in the Tank with Immersed Catcher Electrode (Water temperature: 29°C; Electrical resistance of wtaer: 8000 ohms/cm^3)

Development of Electric Fish Catcher

K.P. Biswas in 2016 had developed an electric fish catching unit for Indian fresh water fishes and tested the catcher successfully.

The experimental unit had three main components, (1) holding tank, (2) DC power source and (3) the catcher electrode. For physiological constants, basic data on the threshold current intensities and behavior of Indian fresh water fishes, tests were carried out in homogeneous electric field in standing and flowing waters

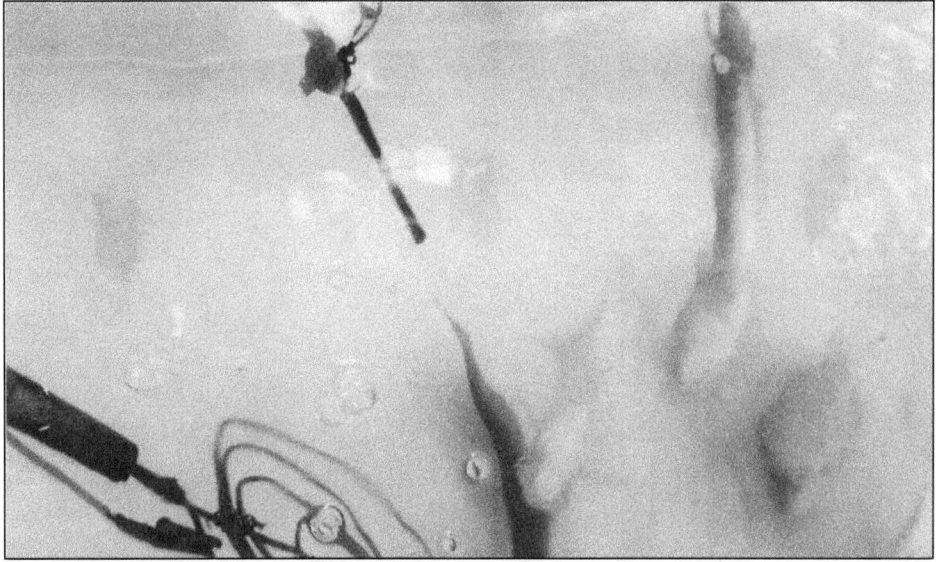

Figure 6: Attraction of Fish to the Catcher Electrode.

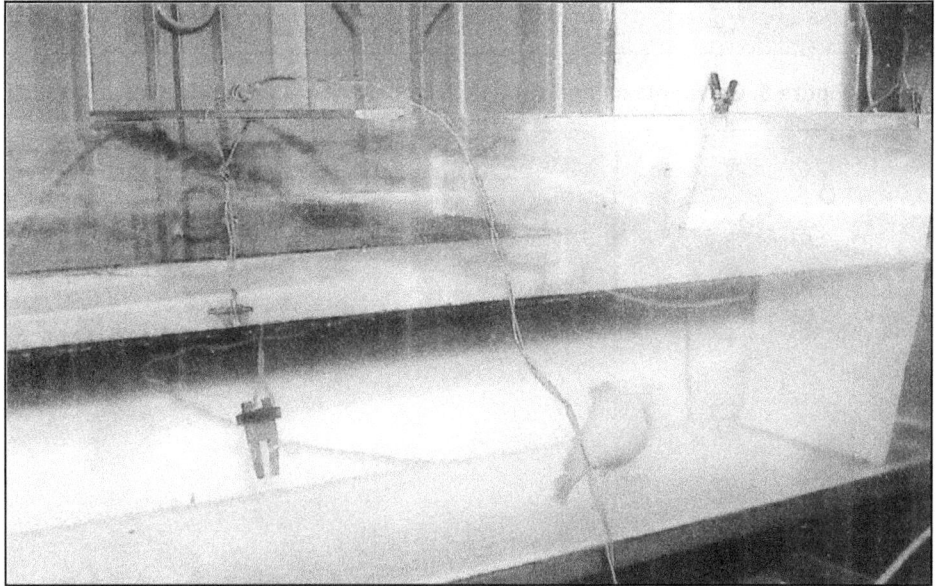

Figure 7: Galvanotaxis of *Notopterus nopterus*.

in continuous direct current (DC) and pulsed direct current (PDC) of very low frequencies.

The data thus obtained formed the basis for the electrical design of a suitable catcher. All experiments were conducted indoors in a glass tank; 1.2 m long, 0.45 m

Figure 8: Galvanotaxis Followed by Narcosis Near Catcher Electrode.

Figure 9: Helic Coil Positive Electrode Influencing Fish to Undergo Taxis.

wide, and 0.45 m deep. The power source consisted of AC from electric supply line through the transformer and rectifier for DC and through electronic low frequency pulser for PDC. The catcher electrode (anode) was a helix coil made of 3 mm copper wire, attached to a wooden handle through a push button switch. In three sets of test where two negative electrodes (aluminum plates of 45 x 40 cm each) positioned vertically at both extremities of experimental tank, 90 to 100 per cent fishes exhibited galvanotaxis and narcosis near the catcher electrode at current densities of 0.00025 to 0.001 micro-ampere per square mm.

Chapter 8

Electrical Fishing in Sea Water

Systematic observations on the effects of electricity on fish date back to the 1930s and the 1950s. In the 1970-1980s European Fisheries Development Organization in The Netherlands, U.K., Belgium, France, Germany carried research and development in the use of electrofishing in marine fisheries, in some cases in collaboration with private companies.

The main motivation for this work was to develop gears which saved fuel particularly during the post 1974 "oil shock" period when the price of oil rose rapidly and electrofishing, which was perceived as being more energy efficient than conventional towed gears, offered the opportunity to save fuel.

However, none of these research programs resulted in a commercially viable fishing gear, largely because it was difficult to make the electrofishing gear robust enough for use in commercial fishing. The method was banned in 1988 in The Netherlands because of fears of increased fishing effort in the beam trawl fleet and development in the other European nations also ceased around that time. European Union Legislation (EU Council Regulation 850/98) banned the use of electricity in 2000.

Since then all legal electric fishing in European waters has taken place under an agreed derogation (that is a permit) from the authorities from these regulations.

Since the 1990s there has been an increased focus on reducing the impacts of trawling, particularly beam trawling, on sea bed habitats. Electrofishing techniques have the potential to reduce the mechanical effects because of the reduced weight of the gear, there being no tickler chains to disturb the sea bed, slower optimum towing speed and to be more selective because larger fish respond more readily to electrical stimulation. This led to a revival of interest in electrofishing and a high level of collaboration between public and private sector in The Netherlands in the development of the "pulse trawl" derived from the beam trawl and the development of the "Hovercran" gear derived from the shrimp beam trawl gear in Belgium.

In a separate development in the early 2000s, it was discovered that razor clams (*Ensis* species) could be induced to emerge from the sea bed through electrical stimulation, rendering them available to collection by divers. Although the method was banned by law, an illegal electrofishing has developed on this species due to its high value.

European Electrofishing Gears

Three main gears have been developed in Europe for marine electrofishing; pulse trawling developed in The Netherlands, designed to replace beam trawling in the southern North Sea, "Hovercran" beam trawl gear designed to trawl for brown shrimp (*Crangon crangon*) developed in Belgium and the illegal razor clam fishing gear, which consisted a towed electrode array which stimulated the razor clams to emerge on the surface of the sediment, after which they were harvested by divers.

"Pulse" Trawling for Flat Fish

The gear used for pulse trawling were originally based on beam trawling gear, but with substantial modification in ground gear. The tickler chains were replaced with electrode arrays and in the case of the HFK Sum Wing with pulse trawl the beam was replaced by a hydro-dynamic wing which was neutrally buoyant, with its position above the sea bed maintained by hydrodynamic forces and via single runner. The Delmeco gear was closer to the conventional beam trawl in design, with shoes supporting the beam at each end. There was a reduction in damaged fish and so quality of the catch was improved (cited in Quirijns *et.al.* 2015). This often meant that although catch rates were reduced in the pulse trawl fishery, profits increased through a combination of fuel savings and increased landing prices.

Hovercran Trawl for Shrimp

The Hovercran gear was developed for use in the brown shrimp fishery with modified ground gear and electrodes. This gear was used in inshore waters of the southern North Sea. The intention behind this gear was to catch shrimps by stimulating them to jump high enough to be captured in the trawl behind and reduce the weight of the foot rope to enable fish and other by catch species to escape. However, in practice it proved necessary to retain some bobbins on the foot rope. Ground contact and by catch was reduced (Verschueren, 2015).

Capture Mechanisms

For marine electrofishing the main aim was to elicit a minimum response to stimulate the fish or shell fish into a position where it could be successfully captured, thus enabling a less invasive fishing technique than was previously necessary. Responses to electrofishing vary between species and size of organism, and this has potential to enable improved size and species selectivity.

Response by Fish

The responses of fish ranged from a "minimum response" consisting of an involuntary contraction of musculature at the make or break of a weak electric field, through to behavior that resulted involuntary swimming, and the extreme response of electro-narcosis at very high field strengths, where the fish cease to respire and sink.

However, behavior varied between species and the size of individual specimens. This was because for a given field strength larger fish was subjected to a greater potential difference from nose to tail or across the body and were likely to exhibit an increased response to a given electrical field. In extreme cases the muscular contraction is sufficient to result in vertebral damage. The wave form of the electric field; direct current (DC), pulsed direct current or alternating current (AC) also has an effect on the fish's response. Experiments carried out in The Netherlands in 1969 (cited in Van Marien, 1997) found that pulsed DC elicited responses at lower field strengths and development of these gears had used this stimulus.

Pulse trawls used electrodes instead of heavy tickler chains to stimulate sole and plaice to elevate off the sea bed and hence became available to the net.

Response by Shell Fish

For brown shrimp, the response consisted of a "tail-flip" reaction which resulted in continuous swimming at an elevation above the substrate of around 10 cm. After around 15 seconds "pulse-fatigue" set in when the animals cease to respond. It was found (Polet, 2004) that the tail flip behavior of the shrimp was different from a range of undersized fish species which tended to stay nearer the sea bed. Hovercran trawl gear aimed to reduce by catch of undersized fish through the use of differing configurations of ground gear designed to enable escape of the undersized fish below the foot rope, whilst the brown shrimp catch was maintained or even increased by electrical stimulation.

The speed of towing was the same as conventional shrimp gear.

Stewart (1974) found that Norway lobsters (*Nephrops norvegicus*) could be induced to leave their burrows if subjected to electrical stimulation. Pulsed electric fields caused muscular contractions, but if these were sufficiently disturbing, the animals took evasive action either by walking slowly out of the field or by using a strong tail flick to make a rapid response ejecting the animals from their burrows by their secondary exits. It was demonstrated that improved catch per effort could be achieved in field studies using an electrified beam trawl in the Moray Firth and the Minch Fisheries (Stewart, 1975a).

Electrical response of bivalves have been reported for razor clams (Woolmer *et al.*, 2011; Murray *et al.*, 2014). It was found that razor clams emerged from the sediment within 37 seconds of the stimulus being applied and reburied within around 7.5 minutes. The illegal electrofishing for razor clams relies on divers to harvest the razor clams once they have emerged from their burrows. Other bivalves and crustaceans were investigated, but did not all show as vigorous response.

Environmental Effects

Replacing the heavy tickler chains on beam trawl gear with an electrode array means that the mechanical foot print of the gear was lighter on the sea bed; reductions in penetration depth for both the Sum Wing trawl with pulse (Van Marlen *et al.*, 2009) and the Delmeco trawl (Depestele *et al.*, 2015) and also the optimal speed of trawling was lower (reduced from approximately 6.5 to 5.0 knots). Therefore from the point of view of reduced mechanical effects on the sea bed and fuel consumption, there were clear advantages of this method.

The fact that larger fish were more stimulated by electric fields means that there was potential for increasing the selectivity of electrofishing. However, it also means that adverse effects due to vertebral damage were likely to occur in larger fish, so large fish such as cod have been examined for adverse effect.

Effect on Cod

Laboratory studies into the effects of pulse fishing on cod discussed in Soetaert, *et.al.* (2015) cited studies that have shown that larger fish were more susceptible to vertebral damage than smaller fish and the closer fish were to the electrodes, the stronger their behavioral response, hence the higher their risk of vertebral damage. Cod exposed to electrical fields of the same strength as used in the pulsed trawl beyond 200 mm from the electrode did not show any injuries, and those beyond 400 mm did not exhibit any reaction to the electrical field. Fish this close to the electrode would be expected to be within the mouth of the trawl and hence be retained. The highest probability of fractures of the spinal column occurred in marketable sized fish and no injuries were observed in cod that were small enough to escape through the meshes of the net. As well as field strength, the nature of the electric field was important; lower frequencies induced more injuries than higher frequencies.

However, two repeats of these experiments by different laboratories failed to replicate these effects, with no injuries observed for cod at these distances from the electrodes (Quirijns *et.al.*, 2015). The differences in effect were ascribed to differences in the body condition of the fish used (*e.g.* Differences in muscular system, mineral content). Some injuries could be induced, but with much higher field strengths than used in commercial gear.

The rates of spinal injury in cod retained in the trawl were examined in an experiment designed to compare pulse trawling with conventional trawling (van Marlen *et.al.*, 2014). Spinal fracture was observed in 4 of the 48 cod in pulse trawl, whilst one spinal haemorrhage was observed in the 48 cod caught in the conventional trawl. Around 2 per cent of whiting in the catch were affected in a similar way. All of these fish were marketable fish retained by the gear. The catch rate of cod per unit area in pulse trawl was 31 per cent of the conventional beam trawl.

The results on spinal damage obtained from the field studies were considered to be valid, but no longer consistent with the laboratory studies, which took place on farmed cod. Further tests on bone and muscle densities would increase understanding of the reasons for the differences.

Effects on other Fish Species

Comparison of laboratory survival rates of plaice and sole captured in conventional beam trawls compared with electro-fished specimens of plaice after 192 hours, but no significant difference in survival for sole. Specimens of laboratory were exposed to electrical stimuli as expected under fishing conditions and examined both internally and externally for lesions immediately after exposure and five days later. No adverse effects could be ascribed to the electrical stimulus (cited in Qirijns *et. al.*, 2015).

Elasmobranch fishes, such as, sharks, dogfish, skates and rays use electro-receptor organs in prey sensing as described in Gardiner, *et. al.* (2014). Dogfish subjected to electric field showed behavioral responses, but no injuries (cited in Quirijna *et. al.*, 2015). However, there is work on progress in Belgium on the effects of exposure of the receptor organs to electric fields, which has not yet been published (Polet, Per. Comm.).

Effects on Benthic Invertebrates

Soetacrt *et.al.* (2015a) and Murray *et.al.* (2014) reported laboratory experiments on a range of benthic organisms, mollusks, echinoderms, crustaceans and polychaetes using behavioral and mortality rates as indicators and most of these organisms showed no significant effects. In a more in depth study Soetaert *et.al.* (2015b) examined the effects of laboratory exposure to electrical pulses in brown shrimp and ragworm. There was no effect of electric field on survival rates. However, microscopic examination of the brown shrimp indicated a raised severity of a natural virus infection in those animals treated with the higher field strength. The authors attributed this differential viral infection to be as a result of the electric stimulation causing stress.

The level of electrical exposure would only occur in a very narrow band along each electrode and for a shorter duration than used in this study.

Trawl Path Mortality

The above description relates to catches retained by the gear. There is clearly scope for fish and other species to be damaged in the gear and not retained. This has been studied for benthic species (Teal *et. al.*, 2014) using sampling from the trawl path post fishing. However, there was high variability so it has proved difficult to detect differences post trawling for these species.

There have been anecdotal reports which suggested that dead fish have been found in the vicinity of pulse trawling. The pathology associated with pulse trawling was well described (above) so given samples in good enough condition, it should be possible to identify whether pulse trawling could be the cause of mortality.

Chemical Effects

The passage of electrical current through sea water will inevitably result in some corrosion of the metal electrodes through electrolysis and there is a potential for the production of hydrogen, oxygen, chlorine and sodium hydroxide. There is also potential for interaction between these materials and the sediments and for other electrical effects on sea bed sediments.

However, this aspects has been very little studied, although video observations by Woolmer *et. al.* (2011) of razor clam (*Ensis*) fishing using a continuous DC stimulus, showed gases being evolved at the anode. Divers participating in razor clam fishing reported that metal diving components can degrade in normal contact with the field, which the diver was not aware of and the electrodes will rapidly erode. Stewart (1974) reported pulses of bubbles being generated at the electrodes when a pulsed DC was used, with the pulse time to concide with DC pulses.

Gases were not observed in laboratory experiments using continuous AC for stimulation of razor clams by Murray *et. al.* (2014). Part of the planned research program in Holland will include a study of the effects of electrical pulses on sediment and geochemistry.

Comparisons between Gears

When a new technology is introduced into a fishery, there are likely to be effects on fishery practices. Therefore initial trials are required to beacted in a structured way to examine the differences between the two gears fished in the same way. This should be followed by a period when the commercial fishery is observed to understand how the changes in technology affect fishing practices. In the introduction of the pulse trawl in the southern North Sea both comparisons have been made.

Comparative Trials

A comparative trial has been carried out by van Marlen *et. al.* (2014) in which the activities of three vessels (one Sum Wing with pulse trawl, one Delmeco "multi-wing" and one conventional beam trawl) were coordinated in an experimental design to examine differences between catches in the electrical and conventional gear.

This study found a 57 per cent reduction in terms of discarded weight per unit area and a reduction of 44 per cent in terms of discarded weight per hour for fish in the pulse gear compared with the conventional beam trawl gear. The difference between two trawls in terms of discard per unit area and per hour reflects the reduced optimal speed of the pulse trawl (5 knots) compared with conventional gear (6.5 knots). The most important improvement was due to reduced discards of flat fish, but also a reduction due to demersal fish as well. There were size selectivity improvements for plaice and sole that were in line with laboratory predictions of Stewart (1975b).

There was a reduction of 80 per cent in surface living benthic discards per unit area and 62 per cent per hour in pulse gear. For infaunal benthos, that is, organisms which live in the sediments, there was a five fold increase of this component in the pulse gear catches; although the absolute quantities caught were small. Whilst benthic species respond to electric field, the differences in catches were considered to be mostly attributed to differences in the way in which the electrode array interacted mechanically with the sea bed and the benthic species compared with conventional gears.

Surveys of Commercial Fishing

Since the introduction of pulse trawl, many vessels in the Dutch flatfish fleet have changed to this type of trawling (Turenhont, 2015) and the pattern of fishing has changed with fishing being concentrated in the southern and western waters of the southern North Sea off Thames estuary. (Rijnsdrop, 2015). Therefore, whilst comparative fishing experiments are vital to obtain a overview of the main differences in environmental effects between the gears, there is a need for more extensive work to monitor and model the effects of this innovation on ecosystems.

Monitoring of the commercial catches of conventional beam trawls and pulse trawls carried out in 2012 by Rasenberg *et.al.* (2013). These results, based on self sampling (where fishermen collect samples) and observer sampling were characterized by high variability indicating uncertainty. There were similar discard percentages (around 50 per cent discard by weight) in the plaice catches in both fisheries and a small reduction in the percentage discard of sole in the pulse fishery. However, the average quantity of plaice caught and discarded per hour was lower; 27-66 kg/hour, in the pulse trawl compared with 87 kg/hour in the beam trawl.

The lack a reduction in discard percentage of plaice in the commercial pulse trawl when the experimental trials indicated improved selectivity (van Marlen, *et.al.* 2014) would require further analysis. It may be due to differences in the patterns of fishing and discarding. The reduction in catches and discards per hour of plaice implies a lower impact of the pulse trawl on the plaice population and was in line with increased targeting of sole by the pulse trawl fishery.

Discards of starfish and crabs were lower in the pulse trawl with pulse trawling catching 16 per cent and 42 per cent respectively of the quantity caught in the conventional beam trawl. This indicates that the commercial implementation of the pulse trawl has successfully reduced the quantity of macro-benthos (starfish, crabs *etc.*) retained in line with the experimental results. There were insufficient cod caught in either gear to make a reliable comparison.

Fisheries Management Implications

There are clearly fishery management challenges arising out of the development of the electrical fishing gear. There has been an important precedent in the East China Sea, where pulse trawling was introduced in the Chinese Penaeid Shrimp Fishery in the early 1990s (Yu, *et.al.*, 2007).

Here the increased efficiencies brought about by the use of electrical gear led to a greatly increased catch which was not properly managed and controlled and resulted in severe overfishing of the resource. Although there were measures to control electrical output and other settings in place, a lack of equipment integrity meant that they could be altered in commercial practice. The unregulated use and misuse of the electrical fishing apparatus negated the advantages of electrical fishing and the use of electrical fishing was banned in 2001. There is a clear need to ensure that the environmental impact of electrofishing is well understood and the management measures are implemented to control the effects and ensure that the fisheries do not suffer similar results to that of the Chinese example cited above.

Key elements are:

☆ Sound fishery management, including knowledge of the dynamics of the stocks and appropriate management.

☆ A good understanding of the effects on species which encounter the gear, whether lethal or non-lethal, captured or not-captured, including reproduction and long term effects.

☆ An understanding of the effects on the marine ecosystem and measures to avoid undesirable effects.

☆ Good technical regulation of the gear, with limits on output characteristics and specifically tuned to avoid undesirable effects. Electricity has the capacity for non-lethal stimulation to enable capture. However, there is the risk of excessive field strength which may result in stress effects or excessive mortality particularly in larger specimens. There is a need to avoid a repetition of the situation seen in the Chinese fishery. When manufacturers and vessel owners found methods to increase field strength and the electrical device became an indiscriminate electrical killing apparatus rather than a stimulus device STECF (2012) note; "—— there is scope for a variety of pulse characteristics within the current legislation, variables of pulse shape, frequency *etc.* can significantly affect the impact on organisms and it is therefore difficult to disentangle and identify the key parameters and their effects".

☆ There is a need to understand the changes in fishery behavior relating to the introduction of the new technology. For example, the pulse trawler men are quoted (Fishing News International, October, 2015) that they can fish more easily off the French and English coasts rather than their traditional beam trawling grounds off the South Holland and Belgium coasts; this was also discussed at the dialogue meeting (Turonhout, 3015 presentation).

Such changes will inevitably result in different patterns of effects between the pulse and non-pulse gears, since different environments will be affected. It is also likely to result in different patterns of competition between fishers, since the above example the UK and French fisheries inside 12 Nm are likely to be in closer proximity to pulse trawling than they were to beaming and so potentially be in closer competition for fish.

Electrofishing has been shown to be commercially viable and more profitable than some conventional methods. However, there is a need to take a strategic view at each method if its potential is to be realized.

Pulse Trawling Research

Pulse trawling has attracted a great deal of criticism in its implementation (Sunday Times, 24 June, 2012) and from French fishermen (CNPMEN, 2015) and it has been discussed at the North Sea Advisory Council. In response the Dutch Government have commissioned a gap analysis (Quirijns *et.al.*, 2015) and set an ongoing research program to fill the gaps in knowledge and set up a stakeholder dialogue group to provide a forum for discussion.

They have a continuing research program with the objective:

☆ To provide scientific basis to assess the consequences of the transition of beam trawling to pulse trawling on the ecosystem (by catch, benthos, ecosystem functioning)

☆ Major strands of the research include;

☆ Laboratory experiments on fish and benthos, to examine the effects of electrofishing.

☆ Field experiments : Effects on the sea bed ecosystem and on species in the trawl path post fishing (partly due to be undertaken in BENTHIS project).

☆ Modelling fleet dynamics and ecosystem functioning.

Control of Electric Field Characteristics

In order to answer the criticism from STECF above, IMARES has set out to define pulse characteristics and fishery management procedures for the pulse trawl, including detailed technical specifications for each vessel held in dossier on board and develop limiter control system to avoid excessive electric fields.

The scheme is in draft form but includes measures which would define and limit the electric field in terms of power and voltage, duration and frequency of the pulse. The system would collect records of the vessel's activities and the voltage discharge of the array that would only accessible by the manufacturer of the gear and the authorities. New statistical codes for reporting pulse trawl activities are proposed within the European Union, to distinguish it from conventional beam trawling.

Hovercran Shrimp Trawling

The Hovercran trawl has the potential to improve species and size selectivity of brown shrimp trawl gear with unpublished IVLO data demonstrating a discard reduction between 50 per cent and 75 per cent (Polet per,com,). However, commercial implementation remains under development (Ludemann and Koschinski, 2014).

Assessment and management of shrimp fisheries and stocks is still in development. Currently there are no catch limits and stock is believed to be growth overfished (ICES 2015a) which means that it would benefit from improved selectivity. ICES (2013) highlight the benefit of better management measures for this fishery. There is a need to ensure that the electric fishing does not have a destabilizing effect on the fishery through the introduction of viable management measures.

Experiments with Pelagic Fishes for Design Criteria for Electrical Fish Harvesting Systems

Field experiments to test the efficiency of a scale electrical harvesting system were conducted off Panama City, Fla. with both captured and wild coastal pelagic fishes. The use of electrical fishing in the sea has lagged considerably behind that of fresh water because of the high conductivity of salt water, which results in extremely low load resistance and therefore very high current and power requirements for generation of significant field strengths. Kreutzer (1964) showed pulsed direct current could be utilized economically to harvest fish in the sea provided that the field voltage gradient and shape, duration, and rate of impulses are suitable. Electrical stimulation produces either fright, taxis, tetanus, or eventually death depending upon the electrical field pulse characteristics (Viber, 1967; Halsband, 1967, Lamarque, 1967).

Success of electrical fishing equipment depends upon use of optimum electrical combinations for inducing fright, taxis, or even tetanus. These various responses have successfully been used to commercially harvest marine animals. The principal applications include an electrical fish pump for hardening menhaden in a purse seine (Kreutzer, 1964), an electrical fish trawl (McRae and French, 1965), and an electrical shrimp trawl (Kilma, 1968; Seidel, 1969).

Test Equipment and Procedure

The pulse generator providing electrical energy to the electrode array had an output capability of 12 kVA at a pulse rate of 50 pulses per second with a peak output voltage of approximately 150 V at a pulse width of 0.8 millisecond. The pulse rate could be varied from 4 to 55 pulses per second, and three different output widths were available with the unit, 0.3, 0.5, and 0.8 millisecond measured at the 10 per cent power points. Pulse rise time was around 0.05 millisecond with a sloped decay. The pulse generator output was designed to operate into load resistance of either 0.05 or 0.2 ohm, since the operational array resistance could not be predicted for all variations in field conditions. At these loads, the output pulse was relatively smooth and undistorted, exhibiting only slight imperfections in the decay portion of the waveform. The waveform was distorted with other array resistances.

The compensating pulse is an important feature of the pulse generator and is designed to significantly reduce both electrode electrolysis and electrolysis of any incidental metal within the electric field, such as a ship's hull. Essentially, the same amount of electrical energy is contained within the envelope of the compensating pulse as is within the main pulse except the compensating pulse is of an opposite polarity. The compensating pulse has no effect on fish reaction, since its amplitude is many times less than the main pulse and is below the threshold level of the fish.

The electrode array and pulse generator were designed to effectively energize a minimum volume of water at least 2 meter in cross section and 4 meter long, and provide a selection of minimum electrical field concentration from 15 to 30 volt per meter. Each electrode of the array consisted of a copper tube frame with copper strips arranged in grid pattern. The strips of copper were 15.2 cm wide (6 inches) with square grid openings of 45.7 cm (18 inches) between strips. It has been experimentally demonstrated that the surface of an electrode can be reduced to approximately 10 per cent of the total area and the surface will function electrically as if it were a solid plate. The electrode design reduced the conducting surface of the electrode to approximately 53 per cent of the total area. So, this grid technique was utilized to allow the fish to be led to and pass through the anode for easier evaluation of their response. The cable connecting the electrode array to the pulse generator was a 12 meter length of 1/0 coaxial conducting cable and represented a total resistance of approximately 0.01 ohm, or a total power loss of 20 per cent in an overall array resistance of 0.05 ohm. Coaxial cable was utilized to eliminate pulse distortion and losses caused by inductance in parallel conductors.

Field strengths listed in volts per meter are averages based on measured electrode to electrode values and separation distance between electrodes rather than in situ field strength measurement, because the density of an electrical field in seawater is not uniform. For ease of measurement, the electrode to electrode voltage was measured at the output of the pulse generator and did not take into account cable and connection losses. Also, due to the hookup restriction in the research vessel's instrumentation room, short lengths of parallel conductors were utilized, resulting in a 40 per cent total cable loss. Therefore the true electrode to electrode voltages and average field strengths are related to measured values as follows :

A. 150 V = 90 V electrode to electrode = 22.5 volt per meter.

B. 120 V = 72 V electrode to electrode = 18.0 volt per meter.

C. 90 V = 54 V electrode to electrode = 13.5 volt per meter.

D. 60 V = 36 V electrode to electrode = 9.0 volt per meter.

The configuration of the electrical field at a pulse generator output of 120 V along with actual measured field strengths (expressed as voltage drops measured across 10 cm) at various positions are shown below.

The measurements are fairly close in value but not exact. The pickup probe was attached to a long pole and the measurements taken from the side of the boat. Because of water current and boat movement, it was difficult to hold the probe parallel to the electric field in exactly each position.

Laboratory tests indicated a field strength of about 15 V per meter was required to properly produce electrotaxis in fish of 10 cm long. Field strengths throughout the volume of water within the electrode envelope could be maintained equal to or greater than the 15 volt per meter requirement.

Based on initial field tests, the general zones of fish response produced by the electrode array are a) effective control, b) possible control, depending on fish size and its orientation, and c) fright zone. The zone of control also extends to the back side of the of the positive electrode.

Test Procedure

The test equipment used in evaluating fish response to an electric field consisted of a deck-mounted pulse generator and an electrode array deployed in the water alongside the vessel. Salinity and temperature ranged from 29.5 to 33.8 ppt and 28.0 to 29.6 degree Celsius respectively.

Two separate groups of experimental fishes were used in the experiments and are referred to as captured fish and wild fish respectively. The first group consisted of 393 Spanish sardines, *Sardinella anchovia*; 397 round scad, *Decapterus punctatus*; 390 scaled sardines *Harengula pensacolae*; 228 Atlantic thread herring, *Opisthonema oglinum*; and 37 Atlantic bumper, *Chloroscombrus chrysurus*. They were attracted by lights at night and caught with a 5-m lift net in the northern Gulf of Mexico and held in a tank of circulated sea water. Prior to testing, each fish was inspected for damage, and only fish in good condition were used. Each fish was exposed to a preselected combination of pulse rates, voltage and pulse widths by carefully dropping them into the electrical field facing toward and within one meter of the negative electrode.

The second group (wild fish) was not handled by the investigators but rather was attracted by lights at night to an area between the electrodes positioned next to the boat. When five or more fish were between the electrodes, they were exposed to preselected combinations of pulse rates, pulse widths and voltage. Visual observations were used to estimate species composition, approximate size, and responses.

To evaluate in situ effectiveness of the pulse characteristics tested, measurement was taken of the per cent of fish which escaped from the electrical field and the per cent which swam the length of the field to he positive electrode. The captured fish were introduced into the field in such a way that they were forced to turn 180 degree in order to swim to the + ve, whereas the wild fish schools were randomly oriented. Fish not electrically stimulated when place between the electrodes exhibited immediate escape movement toward the cathode, the side or down, but usually did not escape by swimming toward the anode since they were dropped into the electrode array facing the cathode. Test fish would occasionally mill between the electrodes. Consequently, the reactions of the electrically stimulated fish were evaluated in terms of electrotaxis or a positive response by their directed behavior to the anode. It is considered swimming to the anode a positive response. All other responses were designated negative.

Results Obtained

Captured Fish

Voltage, pulse width, and rate are equally important for controlling the species tested. Comparison between the pulse widths indicates that a higher percentage of experimental fishes were controlled at the wider pulse widths (0.8 ms). The lower and intermediate stimulation voltages (60 and 90 V) were not as effective in controlling the fishes as the higher voltage (120 V). Furthermore, the combination of 0.8 ms pulse width with 120 V appeared to be adequate for inducing electrotaxis at the widest range of pulse rates (20 to 35 per second).

The ideal pulse rates for inducing electrotaxis varied for each species. Spanish sardines and scaled sardines were under good control at 20 to 35 pulses per second and round scad at 25 to 35 pulses per second. Based on the limited data for thread herring and bumper, the best pulse rates were 25 to 35 pulses per second.

A factorial analysis was used to determine the most effective combination of pulse rate, width, and voltage for controlling Spanish sardines, scaled sardines and round scad. This analysis demonstrates that selection of the proper level of voltage, pulse width, and pulse rate are clearly important for controlling these species. Another important aspect may be the interactions between the three main effects, although the meaning behind this significance is uncertain. These interactions vary between species. Interdependence was observed for all tested electrical combinations in scaled sardines and Spanish sardines. But this was not observed for the round scad. Voltage and pulse rate interact for scaled sardines and round scad.

General observations for the captured group of fish indicated that if the electrical combination was not adequate, these fish would immediately escape to the side or towards the cathode. However, at times when the pulse width was 0.3 ms, thread herring would elicit a jump and skip on the surface of the water and dart out of the field. This escape behavior was never observed at the wider pulse widths. Controlled fish would swam to the anode and circle between the plates of the electrode from inside of the field to the back of the field and back again in a circular swimming motion, and were held until the power was turned off.

The most effective electrical combinations for each species tested are given below.

Effective electrical combinations based on a minimum of 70 per cent eliciting a positive response (captured fish).

Species	Volts	Pulse Rates	Pulse Width
Spanish sardines	90	25-35	0.8
	120	20-35	0.8
Round scad	90	25-35	0.8
	120	25	0.5
	120	25-35	0.8
Scaled sardine	90	25-35	0.8
	120	25	0.5
	120	20-35	0.8
Thread herring	90	25-35	0.8
Bumper	90	25	0.5
	120	25	0.5

It is felt that if 70 per cent or more of the experimental group responded positively, the combination was effective. Output voltages of 60 V or less were ineffective for controlling fish regardless of the pulse rate or pulse width. Effective fish control required an output voltage of at least 90 V with a pulse rate of 25 or more, and except for bumper the pulse width had to be 0.8 ms. An overall effective electrical combination was 120 V at 25 to 35 pulses per second at 0.5 to 0.8 ms, and 90 V at 25 to 35 pulses per second at 0.8 ms.

Wild Fish

The wild fish group was attracted into the electrode configuration by a surface night light positioned above the electrodes and then stimulated. Usually it was successful in attracting sufficient quantities of fish to evaluate a specific combination of electrical parameters. However, their exact position was never the same, especially when a large school of 30 to 50 fish were positioned between the electrodes. Only electrical field characteristics was used, which appeared to be successful during daytime experiments with individual captured fish. The wild fish were only exposed to a pulse width of 0.5 ms, as the time between tests did not permit a change in pulse width. Since this pulse width provided satisfactory results, it was felt that either 0.5 or 0.8 ms would be satisfactory, as indicated from captured fish experiments. Visual observations indicated that the larger fish (more than 10 cm) reacted more quickly and swam to and from the anode before the smaller fish (less than 10 cm) did. From the details of nighttime observations with wild fish, it revealed that Spanish sardines and round scad were controlled adequately at 120 V and a pulse rate of 25 to 35 pulses per second with a pulse width of 0.5 ms. When large schools were attracted between the electrodes, it was not always possible to control all of the fishes. The visual observations indicated that fish in the fringe area would escape since the voltage gradient was insufficient to control fish in the fringe areas. The number of fish escaping probably varied with their position in the electrical field and their size, since smaller fish require higher voltage gradients for control than large fish. At 35 pulses per second and 120 V, it was possible to pull or force fish into the electrode array from the back side of the positive electrode. Positive reactions were elicited in all species at the prime voltage of 120 and pulse rates between 25 and 35. The results from the wild fish experiment conclusively demonstrate that coastal pelagic fish species tested can be controlled and led with combinations of 120 V, 25 to 35 pulses per second and a pulse width of 0.5 ms. The range of pulse widths was slightly narrower in the field than in the laboratory where the experimental tank maintained an uniform field, test fishes could not escape, and in which narrow pulse widths were not possible. Since wide pulse widths require more electrical energy, it is desirable to select the narrowest pulse width possible which will allow proper control of the species. The in situ investigation clearly demonstrates that pulse widths between 0.5 and 0.8 ms can be effectively employed in open water situations in conjunction with proper field strengths and pulse rates.

A review of the field test data suggests an additional parameter, minimum pulse control power for a specific pulse width and field strength should be determined.

The power required to control fish is presented below, based on the parameters of pulse width, pulse rate, and field strength which was used as a criteria during the field tests.

Power for each pulse (*Pp*) is described as:

$$Pp = Ve \, X \, I \, X \, Pw \tag{1}$$

where,

Ve = Electrode voltage

I = Current at load resistance in amperes

Pw = Pulse width in milliseconds

The total load resistance equaled 0.05 ohm with an electrode to electrode resistance of 0.033 ohm and a cable loss of 0.017 ohm. Slight daily variations of 0.006 ohm were noted in electrode to electrode resistance due to small changes in salinity and temperature. For computations, the resistance values was rounded slightly and the electrode to electrode voltage was established as 60 per cent of the output voltage. The current (*I*) and the electrode to electrode voltage (*Ve*) at selected output voltages using an array resistance of 0.03 ohm and a loss resistance of 0.02 ohm were;

Output Voltage	Ve	I
150	90	3000
120	72	2400
90	54	1800

The total power (kW) delivered into the electrode array after cable losses can be computed as follows:

$$Pt = Ve \, X \, I \, X \, Pw \, X \, Pr \tag{2}$$

where,

Pr = Pulse rate, pulses per second

Pt = Total power in kilovolt ampere

Using the above values, the total power for effective electrical control values used was;

Ve	Pr	Pw	Pt
54	25	0.8	1.94
72	25	0.5	2.16
72	25	0.8	2.77
90	35	0.3	2.84

The preceding results suggest there is a minimum requirement of total power (*Pt*) to properly control the fish which would be a constant regardless of the specific combinations of pulse width, pulse rate, and field strengths. Once the effective field strength of 15 V per meter is exceeded, it appears that different minimum values of pulse rate and pulse width can be obtained to produce equally effective fish response. Unfortunately, there are too few data points to support this conclusion. To properly substantiate such hypothesis, it is required to determine either a minimum pulse width for a constant electrode voltage at each pulse rate or a minimum pulse rate for a constant electrode voltage at each pulse width. Without this, it cannot be definitely stated that a parameter of total power (Pt) can be used as a control specification rather than various combinations of electrode voltage, pulse width, and pulse rate. Any more tests would be needed to substantiate the hypothesis, although this approach would be advantageous from a designer's standpoint.

120-kVA Pulse Generator Design

The primary objective in the design of pulse generator was to produce a system which based on the results of the 12-kVA pulse generator electrical fish control experiments would provide the capability for prototype development and effective harvest of fish in several modes of system operation. The output power of the pulse generator and pulse control characteristics were established to satisfy requirements for automatic fish harvesting without nets (Klima, 1970), electrical mid-water and bottom trawling for fish, and to provide the potential for prototype development of possible future applications, such as, fish barriers, electrical aquaculture cages or other such applications.

Netless Fish Harvesting Mode

The initial reason for the development efforts in the field of electrical fishing was to eventually achieve the automatic fish harvesting system. Since this application imposed the most serious power demands, the design specifics were established around that set of conditions and results of the study were used to calculate the power requirements for a netless fishing system. Allowances were made, however, for application of the system to other electrical control applications.

Use of lights at night concentrate fish (Wickham, 1971) in a volume of water which can then be electrified. The minimum volume of water within a light field which needs to be effectively covered electrically to produce commercial quantities of fish would be 5 meter in cross section and 10 meter in length. An equation for resistance of seawater between the electrodes is;

$$R = \frac{pL}{A}$$

where,

L = Distance between electrodes in meters

A = Surface area of the electrodes in square meters

P = Resistivity of sea water in ohm meters

According to this equation the load resistance of two parallel plates is;

$$R = \frac{0.213 \times 10}{25} = 0.0852 \text{ ohm}$$

Where p at 30 ppt and 24 degree Celsius = 0.213 ohm-m

However, this formula only describes the resistance of the volume of water between two electrodes as if the electrode array was a finite conductor. In actual practice, a significant spreading of the electrical field occurs in seawater. If the size of an electrode is small in comparison to the separation distance, the configuration of the electrode in the array is the principal factor in determining the resistance value, as would be the case with small balls or cables for electrodes. For this situation, the size of the electrodes and separation distance are equally important. Since the load resistance of the array in seawater is extremely low, the resistance value used to calculate power requirements becomes extremely important. A small error in the resistance could result in a large miscalculation of the necessary power requirements. For this reason great care has been taken in computing resistance accurately. Resistance measurements for this situation can be calculated by two methods referred to as Kreutzer and empirical technique. Kreutzer developed a formula for calculating spread resistance for one electrode;

$$Rs = \frac{Ko\,(1 + T \times 0.02)}{\text{Root over } A} \tag{4}$$

where,

Rs = Spread resistance of one electrode including field Fringing

Ko = A constant at a specific salinity

T = Temperature in centigrade

A = Area, square meters

The constant Ko varies with different salinity values and must be recalculated for each new salinity. It can be obtain by solving for Ko in equation (4) which requires knowledge of resistance, surface area, and temperature. Once the value of Ko is determined for a specific salinity, equation (4) can be used to calculate Rs for varying electrode surface areas. Because the value of Ko varies with different salinity and is difficult to determine in situ resistance measurements are required, it is decided to establish an empirical ratio which compares the theoretical calculated resistance from equation (3) to an actual measured electrode resistance. The calculated resistance according to equation (3) using the 2 x 2 x 4 m electrodes of one test was;

$$R = \frac{0.189 \times 4}{4} = 0.189 \text{ ohm}$$

With a salinity of 32.9 ppt and a temperature of 28.7 degree Celsius ($p = 0.189$ ohm-m). The measured resistance was actually 0.039 ohm. An index of difference between the calculated and measured resistance provides a ratio of 4.85. The ratio of calculated to measured resistance ranged from 4.85 to 5.2 throughout the study period, with the measured resistance of the electrode array varying from 0.035 to 0.04 ohm. Hence a midrange value of 5.0 seems the most practical and resistance value one-fifth of the equation (3) calculated value is used to compute total spread resistance as shown in the following equation;

$$Rt = \frac{pL}{5A} \tag{5}$$

where,

Rt = Total spread resistance including both electrodes.

As a cross-check to equation (5) it was also computed the spread resistance from equation (4) using a value of Ko derived from the sample test. The measured resistance of the electrode array in seawater was 0.039 ohm. Since each electrode contributes one-half the resistance, the spread resistance for equation (4) is 0.0195 ohm. In addition, since both sides of each electrode in the tests were exposed, the surface area for the equation is twice that of one side. Using these values, Ko is determined to be;

$$0.195 = \frac{Ko\ (1 + 28.7\ X\ 0.02)}{\text{Root over 2 (2) square}}$$

where,

Ko 0.035 ohm-m

For a 5 X 5 X 10 m electrode array using equations (4) and (5), the following load resistances were determined at 28.7 degree Celsius and 32.9 ppt.

Equation (4)

$$Rt = \frac{0.189\ X\ 10}{5\ (5)\ \text{square}} = 0.01512\ \text{ohm,}$$

Equation (5)

$$Rs = \frac{0.035\ (1 + 28.7\ x\ 0.02)}{\text{Square root of 2(5) square}} = 0.00779,$$

where,

Rt = 2Rs = 2(0.00779) = 0.01558

Rt = 2Rs since Rs is the resistance of one electrode.

As can be seen, the value for the load resistance of a 5-square meter by 10-m array compares favorably when determined by two different equations. The higher value of 0.01558 ohm was used in making power calculations since any electrode array will have some additional resistance due to connection losses.

Results from the field study thus provided the following set of basic design specifics for prototype pulse generator for use with attracting lights in a netless fish harvesting application;

1. Minimum field strength - 15 V/m,
2. Pulse rate - 20-30 pulses per second
3. Pulse width - equal or more than 0.5 ms,
4. Array size - 5 x 5 x 10 m,
5. Load resistance of array - 0.01558 ohm.

Using these specifications, it was determined that the output capability of the pulse generator which would satisfy the requirements by the following equation;

$$P = VI \, x \, fl \tag{6}$$

where,

P = Power, watts

V = Output voltage, volts

I = Current, amperes

F = pulse rate, pulses per second

L = Pulse length or width, seconds

To ensure an adequate field strength throughout the electrode array, a value of 20 V/m was chosen for the power calculations. A maximum pulse rate of 50/s and pulse widths of 0.5, 0.75 and 1.0 ms was also selected to give the pulse generator more versatility. Using equation (6), the power requirement is;

V = 20 x 10 = 200 V for 10-m array

$$I = \frac{V}{Rt} = \frac{200}{0.01558} = 12837 \text{ A,}$$

And at 50 pulses per second and 0.75 ms pulse width

P = (200) (12837) (50) (0.75 x 10 to the power – 3)

P = 96278 W

In an applied system, a cable and connection loss will be experienced. Because of the very low load resistances of seawater, a 25 per cent cable loss can easily be expected. Rounding off the requirement to 90 kVA and after allowing for a 25 per cent loss, a pulse generator of 120-kVA was needed to satisfy the system requirements that was established.

As a cross check of the above designed system, the following formula (Kreutzer, 1964) was used to calculate the effective fish control range of one electrode

R = *root over I x L x p divided by G x 2 x pie*

where,

> R = Effective range, meters
>
> I = Current into the water, amperes
>
> L = Length of fish, meters
>
> P = Water resistivity, ohm-meter
>
> G = Body voltage of fish

To determine the effective range of 20 V/m, a value of 1 m was used for the fish length, fish body voltage is 20 V, and the resistivity is again 0.189 ohm-m.

Allowing a 25 per cent cable loss requires a total input voltage of 267 V at a total load resistance of 0,0208 ohm, and the current in the water is found to be;

$$I = \frac{V}{R} = \frac{267}{0.0208} = 12837 \text{ A}$$

Using these values, range ® is found to be;

$$R = \frac{\text{Root over } 12837 \times 1 \times 0.189}{20 \times 2 \times 3.14}$$

$R = 4.40$ m.

Since this value is computed for one electrode, the 20 V/m range of two electrodes will be 8.8 m. In actual practice, however, the range of two electrodes paired together is greater than twice the reach of one, and a 5 x 5 x 10 m array with 20 V/m can be supplied. At the minimum specification of 15 V/m, the calculated reach of one electrode is 5.08 m.

Since the configuration of the electrode array determines array resistance, various combinations of electrode size and separation distance can change the pulse voltage and current requirements. For this reason, a certain degree of flexibility was designed into the netless fish harvesting mode of the pulse generator. The system is capable of delivering up to 1000 V to an electrode array. However, at this voltage the array shape has to be changed to produce a much higher load resistance to maintain the current at a value which is within the 120-kVA rating of the system and the current and voltage carrying capability of various components in the unit. For instance, at 1000 V the electrode array has to have a total resistance of 0.3 ohm.

Mid-Water Trawling Mode

The pulse generator was also designed for application to electrical trawling. This use of the system requires a significantly different configuration than in netless harvesting. Since the pulse generator components are far too large to consider underwater mounting of the system on a trawl, it was necessary to design the unit for operation through a long power cable. The cable transmits the pulsed power from the vessel to the trawl. A cable length of 2200 feet (670.1 m) was chosen to allow trawling to depth of 100 fathoms (182.9 m) with a cable to depth ratio of at least 3 : 1. The length of the cable is important because as it gets longer its direct current resistance increases and therefore either the cable losses become greater or the size of the conductors has to be increased to prevent excessive losses. Since a large power loss is not acceptable, conductor size and the resulting cable diameter eventually become too large and are limiting factors in the total length and therefore the power then can be transmitted down the cable.

The operation of a pulse generator into a long cable requires careful design in order to work. First, the impedances of the pulse generator, cable, and electrode array have to be properly matched through step-up and step-down transformers to accomplish transmission of the pulse down the cable. Unless impedances are properly matched, the pulse will become very distorted or can be totally lost in the cable. Another serious limiting factor in the operation of a pulse generator through a long cable for trawling is the underwater transformers which match the power supply cable to the electrode array. The delivery of significant levels of power, such as 120 kVA, through a single transformer would require a transformer that is quite large and would weigh several hundred pounds to handle the pulse current into an array with a load resistance of 0.05 ohm.

The first intended application of the pulse generator in a trawling mode was with mid-water trawl. The standard mid-water trawl used was a net that opened approximately 9 x 9 m under water. In actual field measurements, it has been found that the net generally opens between 7.5 and 9.0 m in height. Therefore, the pulse generator is required to accomplish effective electrical trawling on a vertical opening of 7.5 to 9.0 m and a horizontal opening of approximately 9.0 m. In the mid-water trawl application, it is expected that the electricity to provide a combination of fright, leading, and some tetanus to aid in harvesting of fish. Past experiments demonstrated that fish generally accumulate in the mouth and forward body of he trawl. Therefore an electrical field applied periodically should force the fish back into the cod end.

Because of component ratings, loading of underwater transformers, and design restrictions, a power of 80 kVA was chosen as the maximum which could be supplied to the electrode array in a mid-water trawling mode. Since the application of 80 kVA through a single transformer is difficult underwater four electrode pairs and four underwater transformers were chosen to match each pair of electrode to cover 9.0 x 9.0 m net. It was found that a reasonable electrode size could be used which would provide a load resistance of 0.2 ohm for each pair and deliver 20 kVA from each transformer. This meant that by connecting the electrode pairs in parallel, each transformer would carry one fourth the current which would be required of a single transformer at the same total output. In addition, the four parallel electrode pairs would represent a total load resistance of 0.05 ohm which could easily be matched to the other impedances of the system.

Within the impedence matching requirements of the pulse generator, cable, and electrode array, and using the maximum output voltage of 2500 V that the unit is capable of supplying in this mode, 450 V can be supplied to each electrode pair through the matching transformers. The surface area of each electrode pair must be adjusted to provide a resistance value of 0.2 ohm. Therefore, the pulse current of this condition is;

$$I = \frac{V}{R} = \frac{450}{0.2} = 2250 \text{ A.}$$

Using equation (7) at 15 V/m, the range of an electrode is;

R = Square root of 2250 x 1 x 0.189 divided by 15 x 2 x 3.14

R = 2.13 m

However, to accomplish at least a fright reaction required 10 V/m or less, depending on fish size. It is felt that a fright reaction, although not as effective as positive control of fish, will accomplish disorientation and therefore harvest of some fish in an electrical trawling mode. Since the temperature of below-surface water will be colder, one can use higher resistivities than 0.189 in calculations as shown in the following calculation for field reach at 100 fathom water depth. In addition, Kreutzer's equation (7) states that the factor in the denominator goes from 2 to 4 as the electrodes are placed in mid-water. Using these values, it has been calculated the maximum variation of values from the surface to 100 fathom in the 10 V/m range of one electrode to be;

Surface :

Salinity 32.9 ppt, temperature 28.7 degree Celsius, p = 0.189

R = Root over 2250 x 1 x 0.189 divided by 10 x 2 x 3.14

R = 2.60 m

At 100 fathom

Salinity 30 ppt, temperature 10 degree Celsius, p = 0.3

R = Root over 2250 x 1 x 0.3 divided by 10 x 4 x 3.14

R = 2.32 m

Again, the range of two electrodes is found to be greater than twice the range of one electrode. In addition, since each electrode pair based on their required size for 0.2 ohm, will be separated by about 1.22 m, field strength adding will occur. Therefore, the effective range of an electrode pair is significantly more than twice the range of one electrode. By installing one polarity electrode on the head rope and the opposite on the foot rope, one should be able to cover a 9 x 9 m area with the weakest part of the field having at least enough strength to frighten fish. One can also use pulse rates higher than 35 per second, which will immobilize fish more rapidly. In addition, it must be remembered that at distances closer to the electrodes, the field strength increases and reaches values which will effectively lead or stun the fish. Because the size of each electrode is relatively small, current densities capable of stunning fish will be found at some minimum distance from the electrodes. This is not desirable for leading fish in a netless harvesting application and is avoided by using large electrodes, but it is very desirable in a trawling mode where the electrodes are inside the body of the net.

Chapter 9

System Components – Electric Fishing Machine

Basically the function of an electric fishing system is to produce an appropriate electrical stimulus in fish near the electrodes to permit easy capture by netting or to cause fish to stay in areas where nets, trawls, or traps can be readily used (Novotny, 1990). Any electric fishing system requires some minimum effective value of current density produced from the electrodes. The minimum value will vary with water conductivity, temperature and target fish. The current level establishes the perception, effective, and danger zones in the electric field surrounding the anode.

The components of an electric fishing system can be classified into six subsystems according to function. These are (1) power supply to provide the electrical energy to the system, (2) power conditioner to condition (or modify) the raw electric energy to meet the requirements of the specific application, (3) instrumentation to provide knowledge of the electrical performance of the system, (4) interconnection systems to safely carry the conditioned power to the electrodes, (5) electrodes to properly couple the conditioned electrical power to the water, and (6) auxiliary equipment to provide the peripheral functions necessary for successful electric fishing (nets, lights, pumps, aerators, rubber gear, *etc.*) (Novotny, 1990).

When fishing with electricity, it is advantageous to produce galvanotaxic current (PDC or DC). DC generators, however, are prohibitively large for boat-based operations. A three-phase AC generator is generally preferred for most boat based electric fishing applications because it is smaller and lighter (for the same power rating) and better suited as a supply source for most power conditioners (Novotny, 1990). AC generators are more flexible in their output parameters than DC generators and, therefore, more adaptable to a wider range of fishing conditions and water conductivities.

Raw, AC-generated electric power is modified via power conditioners. The function of the power conditioner is to provide the appropriate voltage level and wave form (DC, AC, PDC *etc.*) to suit the specific electric fishing application. A major advantage of modern power conditioners (*i.e.*, Coffelt VVP-15, Mark 22, or Smith-Root GPP 7.5) is the flexibility they afford in terms of wave form, voltage level, pulse rate, and duty cycle. This flexibility enables a single electric fishing system to be used in a wide range of applications. Modern electric fishing systems may employ a combination of transformers, rectifiers, filters, and choppers in their power conditioners.

The individual components of the electric fishing system must be electrically interconnected in order to form the complete system. The interconnection system provides the following functions:

1. The main disconnection switch between the power supply and the rest of the system,
2. Circuit protection devices, preferably circuit breakers,
3. Suitable meters and instrumentation,
4. Appropriate safety (dead man) switches, and most importantly,
5. Proper electrical bonding of the cases of all the components to each other and to any metallic parts of the supporting structures.

The bonding ensures that no two external metallic parts of the entire system (including the boat or other support structures) can ever have a potential voltage between them (Novotny, 1990). The interconnection system should be carefully checked by qualified personnel in order to avoid a potentially dangerous situation.

The requirements of an effective electrode system include (a) establishment of a large effective zone while minimizing the perception and danger zones, (b) flexibility to meet variable water conductivities, (c) ability to negotiate weeds, obstructions and current while producing as little physical disturbance as possible, (d) ease of safe assembly (Novotny, 1990). Commonly used electrode configurations that incorporate these principles are Coffelt's Wisconsin Ring, Smith Roots UAA-6 and various sphere anode arrays.

The two basic electrode shapes are spherical and cylindrical. Spherical electrodes have generally superior electrical properties but have many mechanical disadvantages. The most effective electrode arrays combine the positive aspects of both electrode shapes. Cylindrical electrodes, arranged in a circular shape, achives this. The best example of such a design is the commercially available Wisconsin Ring array. This design utilizes the desirable properties of spherical shapes (limited perception zone, no danger zone, large effective zone), while maintaining the advantageous mechanical properties of the cylindrical electrodes (ease of negotiating obstacles, little physical disturbance, and larger overall effective range).

Two guiding principles with electrodes are (1) always use the largest electrodes possible within the limitations imposed by the physical constraints and electrical limits imposed by the generator and electrical control system (Novotny and Priegel.

1974); (2) if possible, mechanically shield the anodes so fish can not come in direct contact with them (Holmes *et al.*, 1990).

Electro Fishing Power Supply – Design Consideration

Electro fishing makes use of electrical fields in the water to stun fish so they can be netted, examined, counted and returned to the water as a part of a fish census. It is important that the fish not be injured in the process. Alternating current in the water will stun the fish wherever the field is sufficiently strong while direct current fields will cause the fish to swim, involuntarily, to the positive electrode where they are stunned when they reach the strong field near the electrode.

The basic source of power is a three-phase, 230 V, 4.5 kVA, 0.9 PF, 180 Hz alternator driven by a petrol engine. An open delta transformer arrangement allows this voltage to be varied from 92 to 460 V. The 3-phase alternating current can be routed directly to the electrodes or can be rectified to provide a DC output. The direct current power at up to 500 V and up to 15 amperes is switched by means of a solid state SCR control. This direct current output to the electrodes can be either on continuously, or it can be delivered as a pulse train output with a frequency range of 5 to 120 Hz and an "on-time" of 10 to 50 per cent of the period.

The great variability of the output is useful in investigating the fish stunning ability of direct and alternating currents as well as combinations of both on various species of fish. The pulsed output makes it possible to use instantaneous power higher than the continuous duty rating of the alternator.

The variation in voltage is used to obtain suitable current levels for a wide range of water conductivities.

Electrical safety was one of the most important design parameters. The generators used in electro fishing provide more than enough voltage and current to electrocute a person.

System Operation

The engine driven alternator, 3-phase output is routed to the transformer panel through a contactor. This contactor opens the circuit in the event one of the safety interlock switches opens indicating an operator on the boat is out of position. The transformer output is switched by a relay in the power control panel such that in one relay position the transformer output is rectified to be used as DC or pulsed DC output while in the other position the transformer output is used without rectification. The pulse-timing circuit and SCR switching circuit control the DC output. The load relay selects whether the DC or sinusoidal AC output is routed to the electrodes and insures that switching from DC to AC does not take place under load.

The cathode consisted of 10 flexible electrodes, each 1.2 m (4 feet) long, mounted five on each side of the boat. The anode was made up of two 91 cm (36-inch) diameter rings mounted approximately 30.4 cm (12 inches) above the water and approximately 2.4 m (8 feet) apart and 3 m (10 feet) ahead of the boat. The anode electrodes themselves consisted of 15.2 cm (6-inch) lengths of 1.6 cm (5/8-inch) diameter stainless steel tubing suspended from the 91 cm (36-inch) diameter rings by 30.4 cm (12 inches) of copper wire. Each anode electrode has an insulating cover which can be moved to expose various lengths of tubing to the water to aid in the control of current in various water conductivity. A total of approximately 30 of these droppers are normally used for the anode.

At the front of the boat there is a deck for the fishermen protected by a guard rail. The automotive type headlights provide illumination for fish gathering. Behind the front deck is the fish holding tank. The transformer and power control panels are located near the center seat. The outboard motor operator's position is at he rear.

For DC operation, the positive output of the power control panel is connected to one or both of the anode rings while the negative output was connected to the cathode on each side of the boat. For AC operation, one phase wire was connected to one anode ring, the second phase wire to the second anode ring, and the third phase wire to the cathode electrodes.

Safety Measures

The electrofishing power supply provides voltages and currents which could be hazardous for the operators. All electrofishing power leads were routed in conduit separate from all other wiring. Low voltage 12 V DC was used for the switching circuits and lights. All metal parts of the boat were carefully bonded to make sure that there will be no voltage between metal parts in the event of an insulation failure.

The alternator output was controlled by a contactor that disconnected the output to the transformer panel when the 12 V DC contactor coil was de-energized.

Only one push button switch, located adjacent to the left front operator, can energize the contactor coil. In order for this coil to be energized or to remain energized, it was necessary that three safety interlock switches be achieved. These switches were located under the front deck floor mats at the left and at the right front operator position and under the seat cushion at the rear outboard motor operator position. The safety interlock switches insured the power would be interrupted if the operators leaves his seat or fall into the water.

Engine-generator Power Source

The power source used for electrofishing was a Model FS 180-1 engine-generator set manufactured by T and J Manufacturing Inc.It consisted of a petrol engine, 3-phase alternator, alternator-rectifier combination and associated basic engine, alternator, and alternator-rectifier controls.

The petrol engine was a 12 hp, 4 cycle, single cylinder engine and was used to drive the alternator and the alternator-rectifier.

The alternator was a 3-phase, 180 Hz, 230 V unit rated at 4.5 kVA at 0.9 PF. The alternator was the power source for electrofishing. A 3-phase, 180 Hz, primary power system was selected so that the power source and any associated transformers would be smaller and lighter than comparable 60 Hz components.

The alternator-rectifier consisted of an automotive type alternator and rectifier with a DC output of 12 V, 55 A. It was used to charge the starting battery and to provide safe, low voltage for control, lighting, and other auxiliary functions where the operator might contact the power supply.

The engine-generator control panel contained conventional components for engine starting and battery charging. In addition, the panel contained input power control connectors, output power connectors, AC and DC ammeters, AC voltmeter, and a power contactor which could disconnect the alternator output from the output connector.

Transformer Panel

The transformer panel provided the transformers and control to convert the input 230 V, 180 Hz, 3-phase power to selected voltage levels. The two transformers in the panel were connected in open delta so that only two transformer cores were required for this 3-phase transformer configuration. This was a weight saving at these power levels and simplified the coil switching for voltage selection. The three output voltage switched in conjunction with the High-Low voltage switch permitted the selection of any of six, 3-phase output voltages.

The input leads were routed through three 25 A ammeters that provided an indication of alternator load and load balance. When the High-Low voltage switch was in the High position, the transformer switching relays were de-energized and each of the transformers was connected in an auto-transformer configuration with output voltage taps at 332, 391, and 460 volts.

When the High-Low voltage switch was in the low position, the transformer switching relays were energized and the windings were paralleled on each transformer and output voltage taps of 92, 161, and 230 V were available.

The 50 turn winding on each transformer provided a nominal 33 V, isolated winding on each transformer. These 33 V windings were connected in open delta to provide 3-phase excitation for the 3-phase bridge circuit providing the isolated DC power supply for the control electronics. The single phase full wave rectifier bridge in the transformer panel was connected to 33 V winding and provided power for the operation of the series connected transformer switching relays. Since most electrofishing was done in relatively low conductivity, the transformer switching relays were energized for lower voltage output as this was the less common condition. The three output switches in conjunction with the High-Low voltage switch select the 3-phase output voltage. With the High-Low voltage switch in the high position and with a 230 V input to the transformer panel, outputs of 322, 391 or 460 V (Nominal : 320, 390, 460 V) can be selected by actuating an appropriate switch. With the High-Low voltage switch in the Low position the outputs can be 92, 161, or 230 V (Nominal : 90, 160, 230 V). The switch interconnection was unique in that if two output switches were actuated inadvertently, only the lower voltage would appear at the output.

An AC voltmeter was connected across one of the phases at the output of the transformer panel to measure phase line to line voltage.

Transformer Winding Specifications

Core material : Z Silectron, 4 mil

Size (in)	I.D.	O.D.	HT
Core	3.23	5.00	1.50
Cased core	3.04	5.21	1.71

Gross core area : 8.47 sq. cm.

Stacking factor : 0.9

Mean path length : 32.9 cm.

Windings : 7 windings

2-AWG 12140 turns each

4-AWG 14105 turns each

1-AWG 1650 turns

Power Control Panel

The power control panel, upper panel, contained relay and semiconductor power devices and control electronics which permitted the AC output of the transformer panel to be routed directly to the electrodes or to be rectified and switched prior to being routed to the electrodes. A rotary selector switch on the front of the power control panel permitted the output to the electrodes to be turned off, to be sinusoidal AC routed directly from the transformer panel, or to be continuous DC or pulsed DC at frequencies of 5, 15, 40, 80, and 120 Hz. Another switch, the duty cycle switch, can select the "on times" of 10, 25, or 50 per cent of any period for the selected frequency. The three meters on the front panel monitor the DC parameters with one 750 V DC meter for the voltage at the output of the 3-phase bridge, the second 750 V DC meter for the actual voltage routed to the electrodes after switching and the 25 A DC meter for the DC current routed to the electrodes.

With the rotary selector switch in the OFF position, the input 3-phase power was routed to the 3-phase bridge rectifier circuit. The 750 V DC meter would indicate the rectified voltage and the input neon lights would be on. Relay K1 would be energized by the DC input voltage that provided base drive for transformer Q1. Relay K1 would be energized and connected the electrodes to the DC portion of the internal circuitry. SCR1 would be on because of the gate drive through resistors R1 and R2. Capacitor C1 would be charged to DC line voltage through SCR1, CR1, and the electrode circuit. SCR2 would be non-conducting since it had no gate drive and its gate to cathode was shunted by resistor R4. Since SCR2 is non-conducting there would be no output and meters M2 and M3 and lights L2 would indicate no output.

When the rotary selector switch was moved to the continuous DC position, gate drive was removed from SCR1 and was routed to SCR2. SCR2 would turn on routing the DC power to the output through CR1 and meter M3. At the instant SCR2 was gated on, capacitor C1 reversed biases SCR1 and turned it off. SCR1 would have its

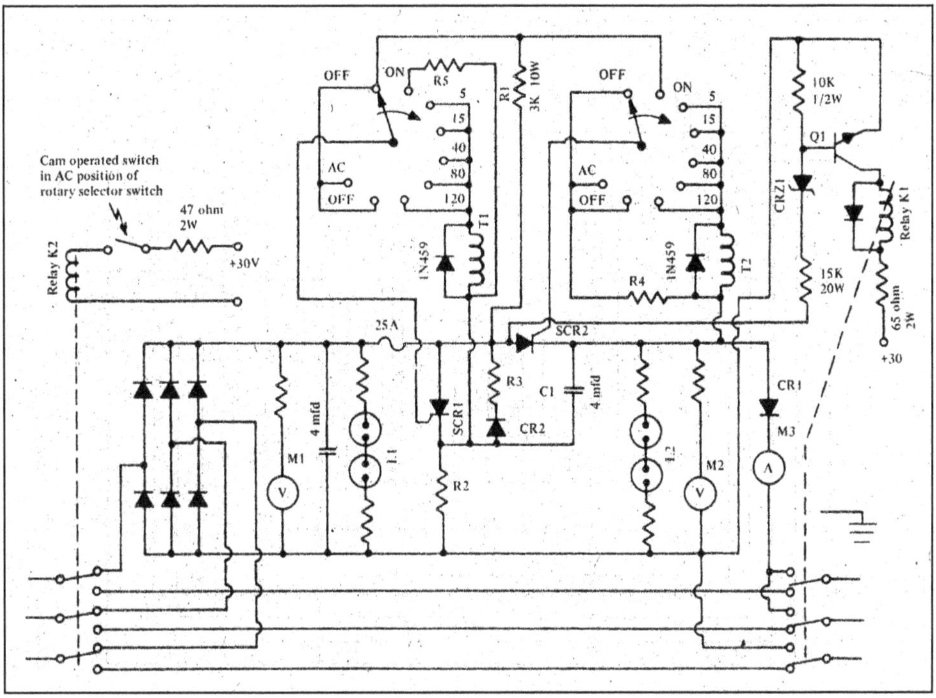

Figure 10: Power Control Panel.

Components: Full Wave, 3 phase bridge, Tung-Sol B-20K, 1000V 35A; SCR-1-2N444; SCR-2N1849B; Q1-TIP29A; R1-3K ohm, 10W; R2-1500 ohms, 200 W; R-310 ohms. 25W; R4, R5-270 ohm, 1W; CR1-600V35A; CR2-600V, 5A; CRZ1-22V, 1W zener; L1, L2-230V neons, 2 in series; T1, T2-Sprague 11Z2000; Relays K1, K2-Potter Brumfield PM17DY 24V: Input Plug-Hubbell 74CM50 with Cover; Output Plug-Hubbell 74CM10 with 7420 Cover; Voltmeters 750V DC: Am-meter 25A DC.

gate to cathode shunted by resistor R5. Rectifier CR1 prevented the 3-phase bridge output ripple voltage from appearing at the electrodes when SCR1 was conducting and SCR2 was off. Without rectifier CR1, there was an AC path from the output of the rectifier bridge through SCR1 and capacitor C1 to the electrodes. With CR1 in the circuit, discharge of capacitor C1 through the electrode circuit was prevented and capacitor C1 was charged to peak ripple voltage to block the current path through the capacitor. Resistor R3 and rectifier CR2 aided in the rapid discharge of capacitor C1 so that it could be recharged with opposite polarity through SCR2 and R2 after SCR1 turned off. This was important when high frequency, low percentage "on time" was desired. The output neon lights were now on and the output meters indicated the output voltage and current.

If the rotary selector switch was returned to the OFF position, SCR1 would be turned on and would provide a path for the voltage on capacitor C1 to turn off SCR2.

With the rotary selector switch in the 5, 15, 40, 80, or 120 Hz position, a switched DC output at the desired frequency was obtained. The percentage "on time" of 10, 25, or 50 per cent could also be selected. In these positions, the SCR pulse transformer, T1, associated with SCR1 provided a pulse when it was desired to turn off the output while SCR2 was turned on by pulse transformer T2 when an output was desired. The pulses from the electronic circuit would turn on SCR1 and SCR2 alternately by action of pulse transformers T1 and T2 at a frequency and per cent "on time" determined by the rotary selector switch. The output neon lights will flash on whenever there was an output pulse and the output DC voltmeter will oscillate as a function of the output voltage peak amplitude, frequency and per cent on time.

When the rotary selector switch was moved to AC, a cam-operated switch energized relay K2 which connected the output of the Transformer Panel directly through the contacts of relay K1 to the electrodes. The control circuit for relay K1 insured that the relay would not switch under DC load since relay K1 would not de-energize until the DC voltage had fallen below 22 V as determined by zener diode CRZ1

The capacitor commutated SCR DC switching circuit was relatively conventional. The high frequency, low per cent "on time" condition made it necessary to make R2, a relatively low resistance, high wattage resistor, 1.5 kilo ohm, 200 W, so that the capacitor C1 could be charged to a sufficient voltage through R2 during the short on time to turn off SCR2 when SCR1 turned on. This same problem led to the use of CR2 and R3 to hasten the discharge of C1 during the short on time of SCR2.

Timing Circuit

A conventional free running multi-vibrator was used to set pulse frequency from 5 to 80 Hz. The multi-vibrator drove a one shot which set the duty cycle. A three-position rotary switch set the duty cycle at 10 per cent, 25 per cent, and 50 per cent for each of the 4 frequencies, 5, 15, 40 and 80 Hz. The one shot drove a buffer amplifier. The output of the buffer was a positive pulse which went through a wafer of the 8 position rotary switch.

The 120 Hz was generated by a fixed frequency multi-vibrator followed by a one shot which set duty cycle by a wafer on the 3-position switch. Pulsed 120 Hz could be obtained at any switch frequency setting when the operator actuated a foot operated switch. S1, on the front deck. The switch energized a relay to switch from the frequency setting of the rotary selector switch to the 120 Hz multi-vibrator. This permitted the operator to obtain the greater stunning effect at any time.

The relay contacts were connected to an emitter follower. The positive pulse from the emitter follower was differentiated by the 0.01 micro-Farad capacitor and 1 kilo-ohm resistor. The positive going pulse derived from the leading edge went through a diode triggering the driver SCR which turned on SCR 2 in the power circuit. The negative going pulse derived from the trailing edge was inverted and triggered the driver SCR which turned on SCR 1 in the power circuit.

The driver SCR's triggered power SCR 1 and SCR 2 through pulse transformers T 1 and T 2. The transformers were necessary for isolation. The pulse to the gates of the power SCR's was about 30 V peak with an exponential delay to 10 per cent in 40 micro-seconds.

Electric fishing gear, based upon the attracting anodic effect of the electric current, has been known in freshwater fishing for many years. In most cases the gear was homemade by fishermen and experimenters, or they have it made by electric engineers to their own design. In some cases, such gear is commercially manufactured.

The first such gear operated with alternating current. It was known, that in 1916 a German, Kramer, used a transportable dynamo coupled with a gasoline motor.

Schonfelder and Schiemenz also carried out fishing experiments in 1916 and 1917 by means of alternating current (220 volts), and with direct current dynamo. Nowadays, in nearly all cases, direct current is used, which as in earlier years was taken from gasoline or diesel generators, or more recently, from batteries. Such gear is used today in nearly all European and in many other countries.

Hand-Crank Magnetos as Fish Stunner

Fishermen often used the magnetos from hand-crank telephones to provide the electrical power to stun fish.

A fish shocker was once a relatively popular way to harvest large numbers of fish from streams, rivers, ponds and lakes.

Fishermen often used the magnetos from the hand crank telephone to provide the electrical power for the shocker, scooping up all the fish they wanted as the stunned fish floated to the surface. The practice is now illegal in many states except when done by licensed fish biologists, who used the shocker to estimate and evaluate fish population under carefully controlled conditions.

Step - 1: Find the positive and negative posts on the rear of the magneto.

Step - 2: Cut the 10 stranded insulated wire into two 15 feet lengths. Strip 2 inches of the insulation off one end of the two 15 feet lengths. Attach the stripped end of one wire to the magneto's negative post and the stripped end of the other wire to the magneto's positive post. Strip about 10 inch of insulation of each wire's trailing end.

Step - 3: Row the boat to an area where fish are likely to school or hide. Drop one wire over one side of the boat and the other wire over the opposite side of the boat. It is best to have at least two people for successful electrofishing (taking fish with a fish shocker).

Step - 4: Crank the magneto, being careful not to touch any area near the rear of the magneto, where electricity is generated. A telephone magneto can generate 8 to 110 volts of electricity depending on the size of the magneto and how fast you can crank. Stop cranking as soon as fish are seen rising to the water surface. The second man should scoop the fish with the net.

Simple Electric Fish Shocker Circuit

This is a simple fish shocker circuit, which helps in catching fish easier. It uses 12 V battery to convert to 220 V AC.

Working - This circuit convert voltage from 12 V battery to 220 volt AC. Both transistors will alternately work on and off, because above phase of transformer has same phase as with below.

The top transistor conduct current before the voltage from 12 V will be sent to the center trap OV to go to the up terminal. It make the center as positive and up terminal as negative and down of 12 V terminal as also negative. Then the negative voltage at below terminal make lead B of all above transistors as bias current.

When the above transistors stop conduct current, magnetic field collapses causing a reverse-phase induction center as negative. The positive terminal is 12 V, then the positive voltage will flow to R to bias lead B of below TR, makes it the switch to connect circuit, causes induction the 12 V to the coil of the lower switch like this forever.

A 12-0-12 volt, 3 A transformer (the general power transformer) was used in the circuit. The transistor 2N3055 was screwed at large heat sink. A resistor R (470 ohms), capacitor C (0.1 micro Farad) and a diode (IN 4007) was connected in the circuit.

The universal board may be used to connect the battery to input and the output to wires into the prongs to shock the fish.

Current will flow from the second terminal of the wire determines the cause of shorting it. It does not shock the user, as it does not complete the circuit, unless the user hold both ends only.

Electric Fishing Gear Types

The following types of gear are used in Sweden (described by Sorensen) :

1. Gasoline operated gear, operating with direct current produced by a generator. The potentials amounts to 220 volts, in some cases to 700 volts. Gear commonly used has a Jap-motor of 1.5 horsepower, 220 volt direct current, weight 65 kg. Catching electrode : 30 x 40 cm. The cathode is a piece of brass cloth of 30 x 100 cm.

2. Battery gear, operating with two or several dry batteries (90 to 120 volts). The gear is carried on the back of the operator and works with direct current of 180 to 360 volts. Good results were obtained with both types of gear in waters of normal conductivity.

In Denmark, according to Larsen, both Swedish and Danish gear has been used. The Swedish (Wolf's electrical fishing gear) consists of a gasoline motor coupled to a generator. It weighs 37.5 kg and produces potentials of 130 to 270 volts. The Danish gear is also operated by a gasoline engine and develops potentials of 220 to 240 volts. It is 47 kg in weight. Good results were achieved with both types of gear, but the waters in which fishing took place may not be more than 2 to 3 meters in depth. Gear producing alternating current has a greater efficiency in larger waters.

But AC often causes breaking of the backbone and damaging of the blood vessel system of the fish.

Much of the electrofishing gear on the market in the Netherlands, as stated by van Drimmelen, is not satisfactory because the water sometimes possesses very high conductivity. For that reason, a special direct current generator of 3000 watts have been designed. The voltage and strength of current, as well as the surfaces of anode and cathode, can be varied and adjusted to the conductivity of the water. Waters of a conductivity of up to 2000 micro-Siemens can be fished with this gear. The electrical fishing technique is used especially in waters of ditches and canals.

As reported by Huet, gear operated by gasoline motor as well as by battery (Salmo-Super) is successfully used in Belgium. Gasoline engine operated is sometimes homemade by fishermen.

Moore reported that electrical fishing gear operating with pulsating current has been developed in the United Kingdom. Energy is supplied by a 12-volt battery, driving a converter which produces an initial potential of 230 volts. From here the current passes over a transformer and rectifier and subsequently over an impulse transmitter. In the end, a potential peak of 500 volts is produced in the water. The gear is 80 kg in weight and supplies 20 pulses per second with a pulse length of 10 milliseconds. It was also used successfully in small lakes.

As stated by Woker, with the exception of two battery gears of the Salmo-Super type, direct current gear operated by gasoline engines is used in Switzerland.

The first commercially-manufactured gasoline generator gear, the *Zurcher Great* (Zurich gear) was produced by Spalti Sohne and Co. Elektromotoren, Zurich. Much of the gear now in use is supplied by Grossenbacher and Co., St. Gallen. The gear consists of three parts:

1. Gasoline generator with carrying frame,
2. Control gear, including built-in cable rolls,
3. Catching electrodes designed as fishing grates

The output of the gear amounts to 1.5 kilowatts. It supplies direct current of maximum 350 volts and load capacity up to 10 amperes. A shunt regulator secures a continuous control of the working potential. Total weight of the gear varies between 85 to 87 kg.

Some 15 to 20 sets of gear are in use, mostly produced by commercial manufacturers of Germany. Electrical fishing is used primarily in fish culture to catch spawning and predatory fish, to control stocks and to investigate mortality. The results are always satisfactory, sometimes excellent, so that the use of the gear is spreading. In addition to the interest in large, heavy electrical fishing gear, there exists also an interest in small portable gear.

Haskell mentioned that in addition to gasoline and diesel generators, portable gears are also used in the United States. Some supply direct current, others alternating current using 6-volt batteries. The direct current is transformed by means of a vibrator into alternating current and the potential is then increased by a transformer.

A selenium rectifier is switched into the terminal circuit for the production of direct current. The potentials produced in this way are between 110 and 220 volts. The weight of the gear is 13.6 kg. When the gear is to be used in waters of greater conductivity, the strength of current is increased by connecting two transformers parallel in the terminal circuit.

An electrical fishing gear to be transported on the back was designed by Philip S. Parker and Edward Burgee of the United States Fish and Wildlife Service. It produced 250 volts of alternating current and weighs 18 kg. The gear can be used in waters with a resistance of 30000 to 150000 ohms per cubic meter.

As described by Burnet, electrical fishing gear has been constructed in New Zealand to operate on the principle of the impulse gear designed by Kreutzer. The gear produces pulsating current through using a condenser discharge. It is driven by a 24-volt battery and consists of a charging and discharging circuit. It produces an approximate potential peak of 700 volts at a pulse rate of 20 per second. The catch, in waters up to 3 meters in depth, amounts to 81 per cent of the stock.

Gasoline Engine Generators

In Germany, several firms manufacture electrical fishing gear. The best known types of gear operating with gasoline generators are :

1. "Mofix" – The direct current generator is coupled to a gasoline motor. Capacity : 1.5 to 3.0 kilowatts; weight including carriage 175 kg.
2. " Sabo" – Portable small Sabo gear producing 0.6 kilowatts and large Sabo gear delivering 1.5 kilowatts, mounted on a two wheel carriage for transport. Both types produce direct current.

Battery Operated Gear

Known types of battery operated gear are:

1. "Bafix" - Supplies direct current; capacity to produce 1.5 kilowatts, weighing 49 kg including 12-volt battery.
2. "Salmo-Super" - Capacity to produce 0.5 kilowatt, producing pulsating current, weighs 40 kg. including 12 volt battery.
3. "Hamburg I - Developed by the *Bundesforschungsanstalt fur Fischeri, Institut fur Kusten-und Binnenfischerei* (Federal Fishery Research Organization, Institute of Coastal and Freshwater Fisheries), Hamburg, manufacturer : Ploger, Hamburg, Kegelhoffstr. 61.
4. Knapsack gear - Capacity up to 0.4 kilowatt, producing pulsating current and weighing 15 kg including 12-volt battery.
5. "Hamburg II" - An impulse gear, connected to gasoline generators, meant for larger waters. Current supply from gasoline generator. The efficiency of the gear depends on the size of the generator. Weight without generator is 15 kg.

Each of these battery types of gear works on a different principle. Bafix operates with direct current, Salmo-Super and the gear designed by the *Institut fur Kusten- und Binnenfischerei* use pulsating currents. The gasoline motor of the Bafix gear has been replaced by a 12-volt battery, which produces the current for the converter. In the Salmo-Super, the current from a 12 volt battery is passed over a vibrator, increased to greater voltage by a transformer and then passed to the electrodes over condensers and rectifiers. The vibrator determines the impulse rate which is about 50 to 60 pulses per second, *i.e.*, the pulses themselves are not governable and changeable. The Salmo-Super produces voltage about 270 volts with a maximum load of about 5 amperes. The weight of the portable gear, including battery, about 40 kg. Peglow provided the Salmo-Super with two vibrators in push-pull connection to have a suitable range of frequency.

The battery gear developed by the Institute of Coastal and Freshwater Fisheries, Hamburg, does not operate with a vibrator, transformer or converter, but with current taken from a 12-volt battery which is passed over a switch shaft specially designed for the purpose. The switch shaft is driven by a small motor. When it turns once around 360 degree, the condensers are switched twice in parallel and twice in series. Forty condensers loaded parallel and discharged in series thus produce a peak voltage of 480 volts. By turning the switch shaft either quickly or slowly the impulse rate (in contrast to the previously mentioned pulse gears) can be altered in any desired way and the gear adjusted to a certain extent to catching the various species and sizes of fish. By switching to several sets of condensers, the impulse period can be adjusted, thus the gear can be adjusted to varying conductivities of water. This impulse gear can also be used in larger waters. For that purpose it must, depending on the area of the water, have correspondingly greater energy. This can be aimed by;

a) A larger size of the gear;

b) Connecting the gear to gasoline generators of 0.6 to 5.0 kilowatts.

Electrodes for Freshwater Electrofishing

For practical reasons, the anodes of nearly all gear described above are metal dip nets with a diameter of about 20 to 30 cm. The anode of Fulda gear has the shape of a metal sieve of about 30 to 40 cm. diameter. In the Salmo-Super the metal dip net can be exchanged for a metal plate, 15 x 30 cm. The cathodes are usually metal plates, 20 x 40 cm. in size. In a few cases, metal rings and metal nets are used.

Experiments made by Kafer to design a metal ring cathode, fastened to the foot of the operator, failed as the space between two electrodes was too small.

The efficiency of electrical fishing gear depends greatly upon the size of the electrodes, especially on their relative sizes. The size of the electrodes is naturally limited by the nature of the waters. They must be handy, because the waters (particularly brooks) are frequently overgrown by shrubs or trees or are difficult to follow and shallow. Practical experience has shown that the ratio of the sizes of the anode and cathode should be abut 2 : 3.

Ranges of Efficiency

The range of all gasoline generator and battery gear described earlier is relatively small, approximately 1 x 1.5 meters. With a favorable conductivity of the water and the bottom and with the occurrence of particularly suitable fish (trout for example) a range up to 3 meters may be expected. These gears cannot be easily used in turbid waters.

The fish are frequently stunned before they reach the electrical hoop net and sink to the bottom and are lost to the fisherman.

The extremes in conductivity of water and bottom in the various freshwater areas often have an adverse influence on the efficiency of electrical fishing gear. In extreme cases, the effect can even completely fade away.

Impulse Generators for Trawl Fishing

The development of useful electrical fishing gear for fishing in sea encountered may more difficulties. The greater conductivity of sea water as compared with fresh water had to be overcome by inboard equipment and the electrical fields and ranges had to be considerably larger. Electrical fishing gear for deep sea fisheries, of course, only of interest when it promises substantially increased catches, and such gear may not require much space aboard, nor complicate the usual fishing methods.

The electro-technical solution of the problem was achieved Kreutzer in co-operation with the *Bundesforschungsanstalt fur Fischeri*, Hamburg, and the Siemens-Schuckert-Werke AG by developing an impulse generator. A condenser which is loaded relatively slowly by a direct current generator is discharged within a very short time over a throttle and pases the concentrated energy over the electrodes into the water to obtain the high peak voltages required. Adjustment of loading and discharging is done bymeans of an electronic governor, whereby the discharge current is switched over two parallel connected ignitrons.

Experiments made in 1952 in the North Sea and the Bay of Kiel with the minesweeper R 96 revealed that it is possible to produce current impulses of at least 10000 amperes. The average capacity amounted to about 180 to 200 kilowatts.

To harness the energy employed for the production of pulsating currents in a practical way, it is necessary to achieve the required density of pulsating current within as large a range as possible, without wasting too much energy in the direct vicinity of the electrodes. According to Haier, that can be reached by various measures;

1. By forming electrodes with largest possible surface;
2. By concentrating the potential at individual points or areas of the field and by giving the current a certain direction by means of auxiliary electrodes;
3. By cumulating the current through short impulses for orienting the field.

The rate of the impulses can be varied, according to the species of fish, from 3 to 60 impulses per second. The length of the impulses can be changed in the ratio of about 1:3.

Use of Generator in Electrical Fishing Gear

These generators can be used in freshwater as well as in sea fisheries, but not advised for freshwater fishing due to their higher price and operating costs. But for sea fisheries they should be more useful. Such a generator can be used in trawl fishery. Fish shoals can be concentrated before the net opening by means of anodic effect, and paralysis prevents the fish escaping the approaching net providing the ratio between the current flow and interruption is correlated to the towing speed. In this way it should be possible to achieve considerable increases in catches.

Such a generator can also be used for driving the fish in a certain direction.

The Russians are also said to have succeeded in paralyzing fish, although according to Skoskiewicz, they have not coupled he generator to a trawl net, but to a fish pump which sucks up the electrically stunned fish.

The gear required for electrical trawl fishing was designed and tested, together with the electrode designed for these experiments. The catching electrode was installed in front of the mouth of the trawl net and the electrical energy transmitted to the electrode by a cable.

Electrodes for Sea Fishing Gear

The electrode designed by Suberkrub, after thorough experiments, has a total surface of 5 square meters. The use of electrodes with large surfaces for freshwater and sea water gear is of importance with regard to the required energy. A function of the shape and surface area of the electrode is the direct current spreading resistance. This decreases approximately corresponding to the square root of the surface.

The opinions on the suitable design of an electrode may vary.

The closed spherical shape of the surface of the electrode is usually impossible because of its unwieldiness and great resistance experienced in towing. A float was used which, owing to its hydrodynamic properties, had the smallest possible weight, the smallest possible resistance to the water, and which took the position opposite the net opening without any difficulties. Perforated bodies can be chosen instead of closed electrical surface. These need not be rigid. They develop their size in the water by the influence of weight, buoyancy or resistance.

Haier thought that it is possible to improve the development and consequently, the effect of the electrical field by using auxiliary electrodes. These auxiliary electrodes should govern the potential and provide the area around them with a greater intermediary potential. This intermediary potential can be produced by either linking the auxiliary electrodes to special potentiometer connections of the generator, or by increasing the most effective conducting or spreading resistance beyond the resistance of the main electrode.

As is known, in the alternating current technique, quickly changeable currents in a medium, through which the conducting cable runs, tend to flow back along that cable. Such an effect of current flow back from the electrodes along the conducting cable, would considerably increase the efficiency of the pulse generator in fisheries as suggested by Kreutzer. In his application for the patent he described;

" In order to keep the potential drop near the catching electrodes as small as possible, the electrodes must be very large. The Ohm resistance of the whole arrangement will thus be reduced. Since only the strength of the electrical current in the water is decisive for the range of the fishing effect, the potential can be reduced in case of a small resistance and consequently also the required capacity. But even here a limit will soon be reached. For a quadruple increase of the surface of the electrode results only a 50 per cent reduction of the spreading resistance of the electrode. One must attempt to restrict the unlimited expansion of the current in the marginal zones of the fishing area, as well as to avoid the great potential drop and thus a high consumption of energy near the catching electrode. The hull of the vessel as cathode is usually large enough not to cause any undesirable losses.

These various advantages are obtained by a special arrangement of the electrodes and the conducting cables. The invention is based on the following observations : If an electrode fastened to a long cable is lowered from a vessel and the cable arranged in the form of a circle, the current expands, if direct current is used, corresponding to the inserted field lines. In this case a wooden vessel with a large electrode beneath the stern was used. The resistance of the whole arrangement is equal to the sum of the resistance of cable and the spreading resistances of the two electrodes. Quite different are the condition with alternating current or with short rated direct current pulses of equal or changing direction. Owing to the large extension of the arrangement, the self induction formed by the cable loop is so great that the inductive resistance becomes large as compared with the resistance of the arrangement in case of direct current.

Unlike direct current, the current does not in this case take the shortest way between the electrodes, but it moves along the cable and turns in that way to the other electrode. The higher the frequency of the employed alternating current, or shorter the direct current pulses used are, the nearer the generated current lines will approach the cable. The Ohm resistance of the water column along the cable passed by the current is, in consequence of the great electrical inductive resistance of the sea water, still small as compared with the inductive resistance of the whole arrangement. The current passing through the cable and that returning along the cable behave like a bifilar winding, the inductiveness of which is approximately zero. The resistance of the scheme is then determined by the sum of the spreading resistances of the two electrodes, the resistance of the cable and the Ohm resistance of the water column. The wide expansion of the current is thereby prevented. The current density along the whole way between the electrodes can be made so great that the whole circuit has a fishing effect. This increases substantially the electrically influenced area. The occurring field is homogeneous in the direction of the cable, and at right angles to the cable, the potential of the field is continuously increasing. When a double loop is formed, like a coil with two spirals, then the current also returns in two loops, so that the bifilar character of the scheme is maintained. The current density in the water is then double that issuing from one electrode. By means of a cable coil with two spirals, the losses near the electrodes are reduced to a quarter as compared with only one loop of the cable for the same voltage in the homogeneous part of the field.

When using two or more loops, it is possible to reduce the space, but yet to meet the requirement, that the inductive resistance of the coil is great enough as compared with the direct current resistance of the arrangement.

The limit for the height of the current pulses is about 20 kilo-amperes. Higher current peaks are not possible, as the normally used ignitrons do not permit any higher peaks. If a current peak of the double amount should be attained, the potential would have to be doubled using the same arrangement.

When using a double loop cable, the potential must be only slightly increased to produce a double current density in the water. The greatest Ohm resistance – at the catching electrode – is in this case only passed by half the amount of current flowing through the water. If it is desired to produce a blocking effect towards the depth, the second cable loop can be put into greater depths than the first one".

If this arrangement of the electrodes is satisfactory, remarkable results can be obtained.

Chapter 10

Factors Affecting the Efficiency of Electrical Fishing Gear

The parameters that regulate electric fishing efficiency are numerous; choice of current (AC, DC, PDC), electrical output, electrode shape and size, turbidity, water conductivity, temperature, depth, habitat, operator's ability, fish species, behavior and size. The most important parameter under the control of the electric gear operator is choice of current. To succeed at electric fishing, one must understand the actual electrical output characteristics (voltage, current, pulse rate, *etc.*) expected in the field. Operators also must understand the widely varying sampling conditions and be able to control current, voltage and pulse shape to properly manipulate the electric fishing equipment, thereby maximizing catch per unit effort (CPUE).

The knowledge of electrical parameters and the components of an electric fishing system must be integrated with the understanding of all the biotic and abiotic external factors affecting catch rate. The most important factors, according to Lamarque (1990) are detailed below:

☆ Water conductivities in fresh waters are divided into three groups. Low conductivity waters, 5 – 30 microsiemens per cm (µs/cm), are represented by mountain streams and lakes or areas associated with high rain runoff. Medium conductivity waters range from 30 – 500 µs/cm; some of the rivers are of medium conductivity range. High conductivity waters have values greater than 500 µs/cm; these are mainly estuaries, brackish water lagoons, and the sea.

☆ Different fishing strategies must be adopted for each conductivity range. Fishing in low conductivity waters is difficult, but good results may be achieved by using very large electrodes (anode diameter more than 60 cm) and high peak voltages (800 – 1650 volts). Best results in medium conductivity waters are achieved with a combination of large anodes and galvanotaxic current. In high conductivity water, PDC (rectangular waves of either 400 Hz or 100 Hz at 10 per cent duty cycle) and smaller electrodes are needed to reduce energy requirements.

☆ Fish behavior in electric fields (electrophysiology) has a measurable effect on CPUE. The physical characteristics of the sampling habitat/fishing ground also play an important role in determining fishing success.

☆ Predatory fish (Percidae, Centrarchidae) are more easily caught than prey species. Spawning or territorial fish are less likely to be frightened out of an area, thus allowing the boat to come in close. Bottom fish and poor swimmers are relatively difficult to catch. Thick scaled fish like carp seem to be more electrically resistant than thin scaled fish such as trout. Many fish build up a tolerance to subsequent electric fishing. Schooling species are easily frightened out of a fisher's effective zone by physical disturbances in the water.

Smaller fish have less body size for a voltage difference to develop across, making them harder to catch than larger fish. Vegetation and cover can hide stunned fish from capture.

☆ Fishing over a gravel substrate produces the best results. Electrode contact with muddy bottoms can short circuit the field, causing a decrease in resistance, which can lead to overloading of the generator. In strong current, tetanized fish often are not visible and, therefore, are washed away from the netters. Turbid water allows a close approach towards fish but reduces catching efficiency through poor visual contact. In general, electric fishing efficiency decreases in moderately fast waters deeper than 3.5 meters.

☆ Electrofishing in an upstream direction is recommended for catching salmonids. Electric fishing should be started at the bottom block-net thoroughly checking for fish that may be impinged on the net. Progress should be made systematically taking care to electrofish all habitat in the stream channel. Complex habitat, such as, debris jams and deep water areas will require more effort than homogeneous habitats. It is often effective for netters to remain downstream from the operator and to keep dip-nets within one meter of the anode.

☆ In homogeneous habitats it is often effective for the electrofisher to move the anode in a W shaped pattern across the stream channel while wading upstream. Netters should be prepared at all times to net stunned fish.

☆ In complex habitats such as debris jams and undercut banks, it is often effective for the operator to insert the uncharged anode into the debris, depress the switch and slowly move the anode into open water areas. Fish will often be "pulled" from the debris into the open water where netters can capture them. Complex area, such as these often conceal several fish and should be thoroughly worked out until no additional fish are captured.

☆ In deep water areas, such as, pools or deep runs, it may be difficult to capture fish. One effective technique may be to "chase" fish into shallow water areas where they can be easily captured. The operator can keep the catcher charged while moving it back and forth across the channel and moving it up and down in the water column. Netters should attempt to capture fish that become stunned but should remain conscious of the water depth to avoid submersion of hands or arms in the water. Slowly moving upstream the entire pool area can be fished electrically. Fish will often flee the deep water moving upstream.

☆ In fast moving water areas, it is often effective for the operator to insert the anode into the water an arms reach upstream, depress the switch, and move the anode downstream approximately in the same velocity the water is traveling.

Netters should have their nets pinned against the substrate in the fast water areas. Constriction points in the stream flow, such as, between two large rocks, make good areas for dip-net placement. The operator can move the anode downstream into the dip-net and then release the switch to discontinue shocking. Netters should check their nets frequently in these areas because fish will often become impinged in them without the knowledge of the crew.

Large streams and rivers become more problematic to catch fish due to their large size. It is practical to catch longer linear stream sections as the sampling reach length is increased and the number of fish sample is sufficiently large, the proportion of fish that potentially move in or out of the site at the margins becomes a small proportion of the population.

Multi-electrode boom arrays have been developed to overcome the problems associated with catching fish from large rivers and canals. The boom mounted, pulsed direct current equipment caught 48.4 per cent of a known population, compared with 24.6 per cent for hand-held gear. Three advantages of boom fishing are low cost, increased maneuverability, and greater catch per unit effort.

To overcome the excessive power demands required to fish high conductivity waters (> 800 mhos), a pulsating direct current control box was developed. It fires up to ten electrodes, energizing one at a time, beginning outward and progressing inward. This sequential firing system presents the electrode array as a single elongated anode with a field of more than nine meters. This system can successfully fish waters with conductivities of more than 4000 mhos. There may be some inadequacies to this system when it is used in very fast, deep and wide rivers.

Pulsed direct currents were the most effective at inducing temporary immobilization of fish. Large fish are generally more susceptible to electric shocks than small ones, because larger fish intercept more current.

Direct current electric fishing worked well for catching night-active bottom-feeders from the stony littoral zone. Ideal conditions were when the weather was calmd light with few shadows or reflections on the water, shallow depth (< 1 M), good water clarity, a homogeneous gravel bottom, gently sloping shore and the lack of vegetation

To determine threshold power densities to tetanize channel catfish over a range of conductivities, pulsed AC and 30 Hz and 120 Hz PDC were used. Fish were exposed to the field for one second. The threshold power densities increased with increasing conductivity. Peak power densities required to tetanize at 100 µs/cm ranged from 4.8 to 13.5 µW/cubic cm. and at 10000 µs/cm ranged from 81.0 to 515 µW per cubic cm.

In order to improve catching efficiency, an electric seine net equipped with pulsed current was developed. The foot rope of a 1200-m drag seine net was bound to an electric wire with 20 per cent of it insulation stripped off to act as an anode. The power output was three phase, 220 V, half wave rectified AC irregularly pulsed at a frequency of 10 Hz. This current was controlled to drive, concentrate and seine the escaping fish.

Fish guiding efficiency tended to decrease with increasing water velocity. The guiding efficiencies of the electrical system at water velocities of 0.2, 0.5, and 0.8 m/s were respectively, 84.2, 54.2, and 50.2 per cent for Chinook; 82.4, 47.8, and 42.8 per cent for coho and 69.9, 40.2, and 44.8 per cent for rainbow.

Factors Affecting the Efficiency Elelectric Fishing

Environmental	Biological	Technical
1. Abiotic	1. Community structure	1. Personnel
a. Conductivity	a. Species diversity	a. Size of the crew
b. Water quality	b. Species composition	b. Experience
c. Water clarity		c. Motivation
2. Habitat	2. Population structure	2. Equipment
a. Habitat structure	a. Density	a. Design
b. Habitat dimensions	b. Size distribution	b. Maintenance
c. Substrate	c. Age structure	
d. Water velocity		
3. Seasonality	3. Species specific	3. Organization
a. Temperature	a. Behavior	a. Site selection
b. Weather	b. Physiology	b. Standard effort
	c. Morphology	

Adapted from Zalewski and Cowx, 1990.

Chapter 11

Electrofishing Induced Injury and Harmful Effects

There are many factors that influence the frequency and degree of injury of fish in electrical fields. Not only there are environmental factors (water conductivity, temperature, physical habitat complexity), but factors such as species caught, unique species behavior, the type of electrofishing equipment used, and wave form produced all have direct influences on potential injury rates. The four basic responses of fish to electric fields are, (1) **Avoidance,** in which fish detect the electrical field and swim away from the anode; (2) **Taxis,** in which fish experience electrotaxis and swim to the anode; (3) **Narcosis,** in which fish are immobilized but muscles are relaxed and respirations are continuing; and (4) **Tetanus,** in which fish are immobilized but muscles are rigid and gills are most often flared without showing respirations. It is in this later response of tetany where the most debilitating injuries occur and most practitioners attempt to regulate their equipment to obtain narcosis avoiding tetany.

The response of fish in electric fields depends upon the field intensity which can be measured by the voltage gradient, current density or power density (Snyder, 2003). Current density and power density cannot be directly measured and must be calculated. As such, they have limited practical applications for field studies. Voltage gradient is the average voltage differential per unit distance along lines of current conduction or flux between two isopotential surfaces and is usually expressed as volts/centimeter (V/cm). Voltage gradient can be physically measured in the water with a voltage gradient probe connected to an oscilloscope or voltage multimeter. According to Kolz (1989), the effectiveness of electrofishing power transfer is maximized when the resistivity of the water equals that of the fish. This rarely occurs and knowledge of peak voltage gradient (V/cm) typically varies by species, age and size, water conductivity and type of current applied for electric fishing (Snyder, 2003). By understanding the voltage gradient biologists can better predict the potential injury of fish in an electric field.

There have been many studies on the potentially harmful effects of AC, DC, and PDC electrofishing on salmonids (Snyder, 2003). In general, AC fields are most injurious to fishes and continuous DC is least injurious (Reynolds, 1996). Hauck (1949) provided one of the earliest and most comprehensive descriptions of electrofishing injuries in fish. These potential effects include cardiac or respiratory failure, spinal or other related internal injuries, stress and fatigue (Reynolds, 1996). Habera *et. al.* (1996) found spinal and hemorrhagic injury rates of about 3 per cent in AC-shocked rainbow trout within southern Appalachian streams. However in other investigations the injury rates were much higher. For instance, in low conductivity Appalachian streams sampled with AC electrofishing, Habera *et. al.* (1999) found large (> 264 mm) brown trout with hemorrhagic and spinal injury rates of 60 per cent and 35 per cent respectively.

Some authors (Dalby *et.al.* 1996) found greater incidences of spinal injury among rainbow trout captured with PDC over DC electrofishing. Reynolds (1996) devised a classification system for assessing hemorrhagic and spinal column damage in comparative studies.

Criteria for Assigning Trauma Scores for Hemorrhages and Spinal Damage in Fish Exposed to Electrofishing Fields

Type of Injury	Score	Criteria
Internal	0	No hemorrhage present
Hemorrhage	1	Mild hemorrhage with one or more wounds in muscle, separate from spinal column
	2	Moderate hemorrhage with one or more wounds located on the spinal column and less than 2 vertebrae
	3	Severe hemorrhage with one or more large wounds on spinal column more than 2 vertebrae
Spinal Damage	0	No spinal damage present
	1	Vertebral compression
	2	Vertebral compression and misalignment
	3	Vertebral fracture of less than 1 vertebrae or complete separation of 2 vertebrae

Shaber and Carothers (1988) found spinal and soft injuries of 44 per cent to 67 per cent in rainbow trout collected by PDC in the Colorado River at Glen Canyon National Recreation Area. These findings raised concern among managers within the National Park Service and the U.S. Bureau of Reclamation regarding use of electrofishing to monitor endangered humpback chub (*Gila cypha*) populations in Grand Canyon National Park (Snyder 2003). These findings also opened a much larger professional debate regarding the efficacy of electrofishing as a sampling tool for some species and resulted in some state and federal agencies imposing operational restrictions on the use of electrofishing (Schill and Beland 1995).

Most internal injuries resulting from electrofishing do not have external signs (Reynolds 1996). However, several investigators have noted the presence of vertical dark brands or banding on trout that have been electrofished (Fredenberg 1992). These brands or bands generally occur in the area of the dorsal fin where the muscle mass is the greatest.

Ainslie *et. al.* (1998) found these brands in juvenile rainbow trout to be associated with internal injuries. In electrofishing experiments conducted at the National Fish Health Research Laboratory, banding in brook trout, green sunfish (*Lepomis cyanellus*) and largemouth bass (*Micropterus salmoides*) has been documented.

In addition to injury due to hemorrhage and fracture, electrofishing can result in physiological and behavioral changes. Gatz *et. al.* (1986) hypothesized that multiple pass or repeated electrofishing exposures might result in measurable effects on growth. To test this hypothesis, he electrofished with 600 V 120 Hz PDC twice within a 1-3 day period for up to seven months and monitored changes in growth rate of trout. He found the percentage of fish with instantaneous growth rates less than average was significantly greater for fish that were electrofished four or more times during the year and warned that growth studies in populations repeatedly electroshocked are likely to underestimate growth. Similar observations have been made in field studies involving multiple-pass depletion sampling on cutthroat trout (Mesa and Schreck 1989). However, in his review of the literature, Synder (2003) concluded that electrofishing caused no overall long-term effects on growth.

Physiological Effects

In addition to physical injuries caused by hemorrhagic trauma and spinal compressions, misalignments and fractures, fishes also may undergo a variety of stress related effects resulting from electrofishing that could have short-term or long-term implications for their health. Neuro-endocrine changes resulting from exposure to electric fields are known to induce changes in blood chemistry and hematology (Bracewell *et. al.* 2004). These changes can include hyperglycemia (Specter and Schreck 1980) and hyperlactemia (Driedzic and Kiceniuk 1976) and changes in plasma cortisol levels (Barton *et. al.* 1986). Other stress related parameters include plasma osmolality, plasma chloride, plasma protein, hematocrit, leucocrit and hemoglobin (Barton and Grosh 1996).

Bracewell *et. al.* (2004) exposed chub (*Leuciscus cephalus*) to simulated PDC electrofishing and found plasma glucose and blood lactate to be significantly higher than that of the controls within 0.5 hours treatment. These levels peaked within 2 hours and returned to baseline levels within 8 hours. Mesa and Schreck (1989) examined physiological changes in cutthroat trout resulting from multi-pass depletion sampling and fish marking with monochord attached fingerling tags. In their combination of field and laboratory studies they found plasma cortisol and blood lactate to increase significantly in cut-throat trout. Multiple shock electrofishing coupled with marking elicited the most severe stress response. Vanderkooi *et. al.* (2001) found suppressed generation of antibody-producing cells (APC) by anterior kidney leukocytes in Chinook salmon (*Oncorhynchus tshawytscha*) exposed to a 300 V 50 Hz PDC electric field.

Figure 11: Brands (Bruises of dark pigmental discolorations) in Rainbow Trout (*Oncorhynchus mykiss*) Caused by Electrofishing. Brands are usually temporary external mainifestations of spinal injury, but injured fish often lack brands. (Photograph provided by and used with permission of W.A. Fredenberg, Creston National Fish Hatchery, Kalispell, Montana; reproduced from Snyder (2003, Figure 2).

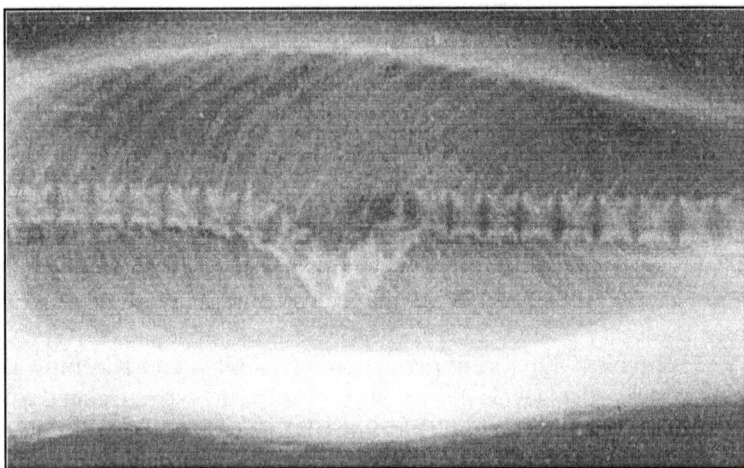

Figure 12: Dorsal-View of a Rainbow Trout (*Oncorhynchus mykiss*) Revealing Severe Spinal Misalignment and Fractured Vertebrae caused by Electrofishing. (Photograph provided by and used with permission of N.G. Sharber, Flagstaff, Arizona; reproduced from Snyder (2003, Figure 16-top).

The authors concluded that electrofishing may have aided the progression of disease in *Renibacterium salmoninarum* infected salmon resulting in the accelerated death of infected fish. It did not, however, make the fish more susceptible to infection or disease. The authors found plasma cortisol and blood glucose levels significantly elevated by 3- hours post shocking and which typically returned to control levels after 24 hours. In some instances, control levels were not obtained until 7-days post

Figure 13: Fractured Vertebrae from a Rainbow Trout (*Oncorhynchus mykiss*) caused by Electrofishing. (Photograph provided by and used with permission W.A. Fredenberg, Creston National Fish Hatchery, Kalispell, Montana; reproduced from Snyder (2003, Figure 18).

shock (Vanderkooi *et. alo.* 2001). Some authors have suggested that full physiological recovery of electro-shocked fishes takes up to 24 hours (Barton and Dwyer 1997).

Implications on Non-target Species

Studies on the effects of electrofishing on non-salmonid fishes are very limited and fewer. Those studies that have been published (Dolan and Miranda 2004) focused mostly on non-salmonid species, that are of interest to fishery managers (*e.g.* largemouth bass, *Micropetrus salmoides*, bluegill, *Lepomis macrochirus*, and channel catfish, *Ictalurus punctatus*). Henry *et. al.* (2004) examined immediate mortality resulting from homogeneous electric fields of 60 Hz PDC on several fishes and found a wide variation in susceptibility among species. The most susceptible species tested was the blackbanded darter (*Percina nigrofasciata*); a relatively large darter found in tributaries to many of the major rivers in the southeast. Recently, Holliman *et.al.* (2003) examined electro-shock induced mortality of the endangered Cape Fear shiner (*Notropis mekistocholas*). In their study they concluded that electrofishing may be safely used to sample small cyprinids when the appropriate waveform is used (PDC, 60 Hz or less). The authors found that mortality was dependent upon waveform, voltage gradient, exposure time and fish length. Kocovsky *et. al.* (1997) investigated the long-term effects of annual electrofishing on stream fish and detected no adverse population effects for brook trout, brown trout or rainbow trout but a notable effect on longnose suckers (*Catostomus catostomus catostomus*). In a study of the effects of both AC and PDC electrofishing on the spotfin chub (*Cyprinella monacha*), Holliman and Reynolds (2003) concluded thet electrofishing should be limited to immobilization to prevent injury and mortality. In their study they found that PDC 60 Hz single-pass electrofishing at 5 V/cm caused 25 per cent mortality in the spotfin chub. This would be unacceptable in any management program for a species-at-risk. In a study of the effects of PDC 3-Hz electrofishing on American eel (*Anguilla rostrata*) in the St. Lawrence River, 60 per cent of the electrofished eels exhibited spinal injuries by radiograph and 30 per cent exhibited hemorrhages (Reynolds and Holliman 2004).Other investigators have found physiological and brhavioral changes resulting from exposure in electrofishing fields. Some species may display reduced swimming stamina (Mitton and Mc Donald 1994), lethargy and cryptic behavior (SWigismondi and Weber 1988), reduced fertility (Muth and Ruppert 1996) or impairment of cardiac function (Scheer *et. al.*2004).

Stress, injuries, and sometimes mortalities among captured fish are unavoidable consequences of electrofishing and most other collection techniques. Among the more effective gear and techniques available for collection of fish, biologists usually select those known to be least harmful, but comparative data on harmful effects are often lacking or inconclusive.

Factors Affecting Injuries and Mortalities

Factors considered in the literature to affect electrofishing injuries and mortalities include type of current, field intensity, duration of exposure, orientation of fish relative to lines of current, and for alternating current and pulsed direct current waveform characteristics, such as, shape, wave or pulse frequency, and pulse width. Additional factors considered were fish species, size, and condition. However, data regarding the effects of these factors are sometimes sparse, difficult to compare, and often questionable.

Available data generally support the contention that of the three types of electrofishing currents, AC is most harmful, DC (constant direct current) least, and PDC usually somewhere between depending on the frequency and complexity of pulses. Although there are reports of no mortality or injury for each type of current, when such adverse effects do occur and comparisons are possible, AC tends to be more lethal than either DC or PDC, and AC and moderate to high-frequency PDCs tend to cause more spinal injuries and hemorrhages than DC, low frequency PDCs, or the only complex PDC tested to date – Complex pulse system (CPS, a patented pulse train of 3 square pulses at 240 Hz delivered 15 times per second). The extent of mortality or injury caused by each of these currents varies considerably with how they are used, other electrical parameters, biological factors, and environmental conditions. With enough field intensity and duration of exposure, any type of current can be lethal, and under certain conditions even DC can injure substantial numbers of fish.

As for most chemical substances and physical parameters affecting living organisms, concentration (in this case, field intensity) and duration of exposure are the primary factors affecting physiological stress and mortality in fish subjected to electrofishing currents. Beyond lethal threshold levels, increases in electrical field intensity or duration of exposure typically result in increased mortality. However, it is not field intensity itself, but the magnitude of voltage differential generated across fish (usually head-to-tail voltage) or specific nerves or tissues that causes electrofishing mortalities and most sub-lethal physiological effects and behavioral responses. Voltage differential is a function of both field intensity and orientation of the fish relative to the lines of current.

Unlike its crucial effect on electrofishing mortality, field intensity beyond requisite threshold levels has an unclear, but evidently not critical effect on electrofishing injuries.

Figure 14: Mass Taxis and Narcosis of Fish in River by an Electrofishing Boat.

Spinal injuries and associated hemorrhages can occur in fish located anywhere in the field at or above the intensity threshold for twitch in the zone of perception. In the zone of perception, as many fish, including those injured by the electrical field, are likely to escape the field as move into the effective zones of the field for capture (taxis, narcosis, and tetany).

The principal cause of spinal injuries appears to be muscular convulsions (myoclonic jerks or seizures) induced by sudden changes in field intensity or, more specifically, in voltage differential across the fish or affected tissues at or above a relatively low threshold in magnitude of change for twitch. Sudden changes occur when the current is switched on and off or pulsed, when fish leap frantically out of and back into the electrified water, and when netted fish are removed from or dipped in and out of the field. Accordingly, duration of exposure in DC should have no effect on incidences of spinal injuries while fish remain in the water, but in PDC, longer exposures subject fish to more pulses and thereby increase potential for spinal injury. However, neither muscular convulsions as the principal cause of spinal injuries in fish nor sudden changes in voltage differential as the principal cause of the convulsions have been experimentally documented. Also, the latter is contradicted, seemingly by the observation of twitches during uninterrupted DC and occasional documentation of as many spinal injuries (at least minor ones) in DC with just two sudden change events (when the current switched on and later off) as in some simple or complex PDCs with numerous sudden changes in voltage differential.

Increase in spinal injuries with exposure time might be expected as well for AC with its cyclic changes in voltage differential and direction (effectively alternating half-sine pulses), but limited experimental evidence suggests otherwise. Perhaps the changes in AC voltage are not sufficiently sudden (if so, the same would apply to half-sine PDC), or the change in direction precludes possible consecutive-pulse summation effects that might sometimes be necessary to achieve the threshold magnitude of change in voltage differential.

Whether the probability or degree of spinal injuries and hemorrhages increases with field intensity or not, fish in a state of narcosis (petitmal) or tetany (grand mal) may no longer be subject to the sudden convulsions that are believed to cause most spinal injuries in PDC (and possibly AC). Injuries might still occur during transition between these states and when fish are removed from the field. If some spinal injuries do occur during tetany, as has long been suspected but unproven, the sustained muscular tension would have to be sufficiently strong to permanently compress one or more portions of the spinal column, burst blood vessels, and possibly fracture vertebrae. Aside from this possibility, measures to specifically reduce the intense zone of tetany around an electrode might not have much impact on the frequency of spinal injuries, but they should reduce incidences of severe stress, fatigue, and mortality.

Orientation of fish when first exposed to the effective portion of the field is probably as significant a factor in electrofishing injuries as in other responses and mortality. However, based on limited evidence, greatest effect appears to occur when fish are perpendicular to rather than parallel to the lines of current (minimum rather than maximum head-to-tail voltage differential). If so, experiments to assess the injurious effects of electric currents on fish might be confounded or biased to minimum effects if fish are held parallel to the direction of current.

For PDC, pulse frequency appears to be a primary factor affecting the incidence of spinal injuries and may be a significant secondary factor in electrofishing mortalities. As expected if spinal injuries are caused primarily by sudden changes in electrical potential, the incidence of injuries is generally lowest for low frequency currents and increases with pulse frequency. With regard to incidences of spinal injuries, the CPS pulse train with a primary frequency of 15 Hz appears comparable to simple low-frequency currents (and DC). It is unknown whether other pulse trains or complex variations of PDC also result in as few injuries as low-frequency PDCs.

The effects of pulse shape or waveform, pulse width or duty cycle, and voltage spikes on mortality and spinal injuries have been inadequately investigated and data that are available are difficult to compare and sometimes contradictory. Although exponential and half-sine PDCs have been implicated as particularly lethal and half-sine, quarter-sine, and square PDCs as particularly injurious, the effects of PDC waveforms on electrofishing mortality and injury remain inconclusive. Likewise for AC waveforms, despite one comparison of sine-wave and triangular-wave AC which revealed no significant differences in incidence of externally obvious injuries but notable differences in the nature and perhaps severity of those injuries. The little data that exists with regard to pulse duration or duty cycle suggests no effect on mortality and a tendency for fewer spinal injuries using currents with longer pulses or greater duty cycles. A limited-scope investigation suggested that voltage spikes have little or no impact on electrofishing injuries or mortality.

Evidence to date strongly indicates that trout, char, and salmon are most susceptible to spinal injuries, associated hemorrhages, and probably mortality during electrofishing than most other fishes. Among other species, burbot (*Lota lota*) and sculpins (Cottidae) were reported to be particularly susceptible to electrofishing mortality, at least under some environmental conditions, whereas goldeye (*Hiodon alosoides*), some suckers (Catostomidae), channel catfish (*Ictalurus punctatus*), largemouth bass (*Micropterus salmoides*), walleye (*Stizostedion vitreum*) and paddlefish (*Polyodon spathula*) were reported to be more susceptible to electrofishing-induced spinal injuries and associated hemorrhages. Electrofished mountain whitefish (*Prosopium williamsoni*) have been reported to be particularly susceptible to bleeding of the gills.

Because voltage differential across fish or specific tissues increases with size, larger fish have been expected to be more susceptible to electrofishing mortality and injury than smaller fish.

Some data support an increased frequency of spinal injuries as fish size increases, but other data do not, and so the importance of size remains questionable.

The physical condition of fish can affect their susceptibility to electrofishing injury and mortality, but assessment of this factor is based mostly on suppositions and casual observations rather than specific experiments and data. Fish in poor health may respond less strongly to electric fields, thereby reducing chances for spinal injury, but they also may be less able to withstand the stresses of tetany and apnea during narcosis, thereby increasing probability of death. On the other hand, weakened skeletal systems probably make fish especially susceptible to spinal injuries. Temperate fishes electrofished during the late fall through early spring may be less likely to suffer either spinal injuries or mortality due to the low water temperatures that substantially reduce metabolism and slow responses.

If there are significant harmful impacts on fish resulting from single electrofishing events, the effects of multiple events should be cumulative. In at least some cases, the stress of repeated handling has greater impact on delayed mortality than repeated exposures to electric fields. The incidence of total injuries among captured fishes inhabiting repeatedly sampled waters increases cumulatively, not only during multiple-pass sessions, but in successive seasons or years of sampling. Some newly captured fish may have been injured during prior treatments or sampling but at that time either escaped the effective portion of the electric field or were missed by the netters.

Impacts on Reproduction, Embryos, and Larvae

Electrofishing can also affect reproduction and early life stages. In addition to or as a result of injuries, exposure of ripe fish to electrofishing fields can cause significant damage to, or premature expulsion of gametes and sometimes reduces viability of subsequently fertilized eggs. Electrofishing over active spawning grounds can also significantly affect survival of embryos on or in the substrate if exposed during their more sensitive stages (prior to acquisition of eye pigment). Exposure of recently hatched larvae might not cause significant mortality but can reduce growth rates for at least a few weeks. Field intensity and duration of exposure appear to be the most critical electrical factors affecting embryos and larvae.

Biswas (1977) experimented on the developing fertilized eggs of carps (*Labeo rohita, Catla catla, Cirrhinus mrigala and Labeo fimbriatus*) by exposing them to underwater AC and DC shocks.

The severity of electric shocks on the developing embryos depended not only on the intensity and the period of exposure, but also on the developmental conditions of the embryos. Besides the developmental and phenotypic anomalies of the embryos, the percentage of mortalities were more in early cleavage stage in case of *C. mrigala* and in near to hatching stage in case of *L. fimbriatus*.

The losses of developing eggs by rupture of embryos along the extremities during DC exposure and premature breaking of the egg shell, rupture of embryos, appearance of white spot on the yolk and abnormal development of embryos, like tail defects as an after effect of current exposure corroborate the observations of Hayes (1949), Godfrey (1957) and Hinrichs (1925). These effects were more intense when the developing eggs were treated with AC indicating the severity of AC shocks over DC.

Among the phenotypic abnormalities of the embryos which survived from the shock treatment, tail defects of *L. fimbriatus* eggs with DC shocks, bend and undeveloped notochord of *L. rohita* and *C. mrigala* by AC stimulation confirmed the severity of AC shocks over DC exposure and was identical to the effect of ultraviolet stimulation on *Fundulus* eggs (Hinrichs, 1925).

The response of *L. rohita, L. fimbriatus* and *C. mrigala* larvae at their three stages of development, *e.g.* newly hatched embryo with yolk, young larvae with fully absorbed yolk and young adult in a DC field of slowly rising field intensity revealed that the hatchlings with undeveloped sensory and locomotor organs could feel the presence of underwater electric field by exhibiting contortion of their bodies, abrupt disorganized swimming and the cessation of voluntary movement. With the advancement in growth, the larvae assembled near the negative electrode with their body axis at right angle to the current lines. At higher intensities they underwent narcosis near the positive electrode. Young adults (20-25 mm) responded by retarded swimming during fright reflex, forced swimming towards the positive electrode and narcosis at higher field intensities, characteristics of a full grown fish. These differences in behavior can be attributed to undeveloped encephalon and afferent paths in newly hatched embryos; since according to Halsband (1955), the galvanotropic response of a directional character is initiated by the reflex center, located at the mesencephalon-cerebellum level, and also by the medullary sensory path way. With the development of the neuro-muscular system of the larva with an advancement of growth, the galvanotropic response of the organism became more and more prominent.

The severity of AC shocks exhibited tetanic condition of the embryos followed by spinal curvature and death irrespective of the species.

Chapter 12

Hazards of Electrofishing: Precautions and Safety Measures

Safety should be a primary consideration in all electric fishing operations. All personnel involved in electrofishing operations should be instructed as to the fundamentals of electricity, and understand and observe the safety requirements associated with electrofishing.

The single most important factor in both electrofishing efficiency and safety is the training and experience of the crew. Regardless of the safety precautions given, the capability of the crew in adhering to those guidelines and good common sense in handling unforeseen circumstances, is of cardinal importance (Smith 1989). It is recommended that crew leaders attend a course on " Principals and Techniques of Electrofishing " either in the U.S. Fish and Wildlife Services Fisheries Academy or in State Institute of Fisheries, North Rhine, Albaum, Saurland, Germany.

A standard set of safety practices are listed below along with two daily field check lists concerning boat and electric fishing equipment. Safety practices should include the following (adapted from H.G. Lazauski and Malvestuto, 1990).

1. All United States Coast Guard safety equipment for the operation of a 28 feet boat should be used.
2. Red Cross first aid and CRP training should be provided for all members of the electric fishing boat crew.
3. All members of the crew should be familiar with the electrical system of the boat.
4. All dip netters should wear rubber gloves, rubber boots, life vests and noise arresters if needed.
5. Boat operators should wear life vests, rubber boots and noise arresters if needed.

6. Electric fishing run should be kept under one hour to avoid netter's fatigue.

7. A strict check, via checklists, should be made of all electrical systems before each day's work in the field.

8. All fishing should cease at the first sign of lighting, rain, high winds, or dip netter's fatigue.

9. Alcohol should never be allowed on an electric fishing boat.

10. Never touch the water or an electrode while the current is on.

11. Refuel the generator after engine has sufficiently cooled.

12. The boat driver should not make sudden turns or changes in boat speed.

13. No unauthorized passengers should ever be allowed on an electric fishing operation.

14. Know the range of your electric field. Avoid public recreation areas. Do not electrofish near people or animals.

15. Avoid all unprofessional conduct (horse play).

16. Carry appropriate spare equipment for the boat.

17. Carry a first aid kit.

Check lists should be developed for all phases of electric fishing operations. These should include items that are used daily, such as boat launching and electrical connections. A sample of electric fishing boat unit inspection sheet is given in Table below:

**Daily Check Sheet for Electric Fishing Boat Safety Inspection
(Adapted from Goodchild 1990)**

Boat # —————————————— Date ——————————

Crew leader ————————————— Time ——————————

Crew members ————————————————————————

Location ———————————————————————————

Log Book up to date Y/N

Manual present Y/N

	BOAT	
— Hull integrit		— Auxiliary motor present and working
— Safety railings intact and sturdy		— Oars/paddles present
— Decks clean, free of water/bilges dry		— Anchors/bailers present
— Adequate protection of wiring		— Controls and gauges operational
— Adequate connectors and interlocking		— hv output checks done
— All metal equipment in boat electrically bonded to hull (check with volt/ohm meter)		— Audible tone generator working
		— bv flash light working

— Batteries fully charged-properly enclosed and vented	— All foot switches working
	— "KILL SWITCH" working
— Communication gear working	— Operators safety switch working
	— Boat clean-equipment neatly stored

GENERATORS/ALTERNATORS

— Unit electrically bonded/connected to hull	— Oil level O.K.
— Exhaust directed away from operator	— Gas topped off
— All electrical connections secure and protected	

BOAT MOTOR

Inboard

— Oil level O.K.	— Auxiliary motor working
— Components secure	— Bilge blower working
— Belts O.K.	
— Visual inspection O.K.	

Outboard

— Proper venting of exhaust	— Fastened securely- safety chain
— No gas leaks	— Adequate gas supply

ANCILIARY EQUIPMENT

— Fire extinguisher present – fully charged	— Communication gear working
— First aid kit and flash light present	— Lights working

PERSONNEL/CREW MEMBERS

— Each crew member briefed on boat	— Crew wearing protective hearing gear
Operations	— Each crew member has a dead man switch
— Minimum number of crew trained in	— Safety procedures covered
CPR and basic electronics	— Local arrangements covered *i.e.* police
— Crew wearing PFD's	— Hospital route outlined
— Crew wearing rubber gloves (long arm)	
— Crew wearing rubber boots	

A detailed instruction guide or manual should accompany each electric fishing apparatus to assist the operator. The operator should be familiar with both the unit and manual before fishing begins. A log book should also be available to record dates and times of use, maintenance, problems, and repairs.

An important emergency procedure is to have a pre-determined plan in the event of an accident. A documented route to medical facilities and procedures to follow is essential.

These safety procedures should be adhered to by all project personnel at all times. The safety check list and log book should be filled out every day. Also, all operational parameters (control box settings and meter readings) should be recorded with field data and any observations of abnormal appearance, behavior, or mortality. This data will help refine parameters for future trips, avoid undesirable effects, and add to the data base on such effects. All members of the fishing crew should be familiar with the checklist material and compliance procedures.

Chapter 13

Electrofishing Policy and Code of Practice

Electrofishing has become an essential sampling tool in the study of freshwater fish ecology. Although used by research groups in Australia for over 30 years, there has been no national code of practice to ensure safe operation and certification of equipment. Several states have had their own guidelines, but increasing use throughout the country means that the time has come for a consistent and coordinated approach to training and safety.

It is the policy of the Alberta Fisheries Management Division (AFMD), Government of Alberta that all electrofishing conducted in waters of the province of Alberta will conform with standards adopted to minimize injury to fish.

The AFMB requires that electrofishing be conducted only when other less harmful means of population sampling will not yield the desired results, and only when careful consideration has been given to the effects of electrofishing on fish populations. The increased attention given to Alberta's fishery resources by both the public and private sector has led to the increased use of electrofishing as a sampling tool. Every electrofishing operation should be preceded by an objective analysis of the necessity of conducting the work, and whether or not the potential benefits outweigh the costs, financially, as well as from a resource impact viewpoint.

Electrofishing of spawning runs and over spawning fish will not be allowed except in instances where eggs or sperm must be collected for government hatcheries or the data being gathered are critical to the well-being of that fish population. Electrofishing over trout redds which contain developing embryos should not occur under any circumstances. Research on impact to eggs and larvae of non-salmonid species also needs to be conducted.

Electrofishing of threatened species may only be conducted after documentation that the type of electrofishing and the method used to apply it will result in the minimum rate of injury to that species.

The table below is given as a guide in selecting equipment and operational settings to minimize injury to fish. Most of the information summarized in the table is derived from a study on electrofishing injury to trout. In general, the guidelines require a high level of training for crew leaders and the use of common sense in the application of electricity to fish.

Patameter	Recommend	Avoid
Pulse rate	30 Hz or less	Over 30 Hz
Pulse duration	5 milliseconds	10 milliseconds or greater
Pulse shape	Smooth DC-Best	Rectified Sine
	CPS- Second choice	Capacitor discharge
	Square- 3rd choice	AC
Voltage	High conductivity = Use low voltage	
	Low conductivity = Use high voltage	
Shocker box	Coffelt Mark 22 M	Coffelt VVP2C
	Coffelt Mark 22 CPS	Coffelt VVP2E
	Coffelt VVP 15 (smooth DC or low pulse rates)	Leach/Fisher pulse
	Leach/Fisher (smooth DC only)	Smith Root VI-A
Generator	Low conductivity (< 200 μmhos/cm) 2500 W or > High conductivity (>200 μmhos/cm)	Inadequate power plant
Electrode	Bigger is better – Use largest possible anode except in highest conductivity water (800 μmhos/cm or more)	Small point anodes such as a single dropper
	Always maximize cathode size In metal boxes use the boat	Never use small cathode
Method	Mobile anode – Best	Never allow fish to lie in field
Intensity	Turn power down to the lowest level	Overkill
Brands	Look for brands, if numerous, turn power down	Branded fish, an indicator of spinal injury
Fish species	Most susceptible to spinal Injury- trout, unknown susceptibility warm water spp.	Never assume fish not injured based on external appearance
Fish size	Be more cautious with large fish	Do not assume small fish are immune to spinal injury
Environmental	Record water temperature and conductivity and adjust methods accordingly	Do not ignore water condition
Eggs	Avoid shocking spawning female and areas with with reeds	Assume eggs in reeds have potential to be damaged
Crew	Use the best trained crew available. Avoid multiple-dipping into field and other factors that will stress fish	Untrained crews are not permitted. Never electrofish under conditions that are marginal or hazardous for you or the fish.

Australian code of electrofishing practice resulted from the international Workshop on Development in Electrofishing conducted in Canberra, March 1995, by the Cooperative Center for Freshwater Ecology and was prepared under the auspices of the Fishery Management sub-committee of the Standing Committee for Fisheries and Aquaculture (SCFFA).

The objective of the code was to ensure safe operation of electrofishing equipment for which;

1. The apparatus must conform to a national standard in design and construction, and be maintained and inspected by qualified electrical mechanics and electronic technicians.

2. Electrofishing personnel must be certified free of major heart or respiratory complaints by a medical practitioner; be trained in the fundamentals of electricity, correct and safe electrofishing procedures and first aid including cardio-pulmonary resuscitation (CPR).

3. All electrofishing operations must follow the standard safety guidelines ontlined below.

Hazards of Electrofishing

Electrofishing equipment uses voltages and currents that can be lethal to humans. Direct effects of electric shock may include heart failure, respiratory interference or electrical burns. There may also be indirect injuries through a worker recoiling violently and striking an object.

The main sources of potential risk of electric shock during electrofishing operations are:

1. Bodily contact with energized electrodes of opposite polarity.

2. Bodily contact with water within the electric field.

3. Shocks from inadequately constructed or insulated equipment.

Other hazards include drowning, fire, exhaust gases, and the increased risk of accident due to fatigue.

It is easy for experienced operators to become complacent about the dangers involved in electrofishing, but they are always present and must be constantly guarded against by regular checks and maintenance to minimize the risk of accidents.

Operators and Crew Training and Certification

All electrofishing operations must be carried out under the supervision and control of a Senior Operator who has been awarded a Certificate of Competency in Electrofishing Procedures and Safety for the particular type of equipment being used *i.e.* backpack, shore-based or boat-mounted.

Initially, certificates will be awarded to persons who can provide evidence of having experience using a particular electrofishing gear type for at least 20 sessaions (an electrofishing "session" refers to a single complete electrofishing operation consisting of gear set up, safety checks, sampling, and gear dismantling, In this way, several sessions may occur in one day if more than one site is sampled).

The evidence of experience may consist of a signed statement from a university department, research section or commercial licence authority, plus a current Senior First-aid Certificate or equivalent including CPR, and a recent medical certificate stating freedom from major heart or respiratory complaints that the examining doctor considers could pose a significant risk should even a minor electrical shock occur. Certificate holders will be provided with a manual outlining electrofishing theory, procedures and safety which they will be required to follow.

The onus will then be on these Senior Operators to train and supervise other staff and students under their control, complying strictly with the requirements outlined in the above manual. All electrofishing team trainees must have also completed a first aid certificate and pass a medical examination, and are required to keep a log of their hours using each gear type. On reaching 50 hours, the supervisor will sign and forward the log to the National Electrofishing Safety and Training Administrator, who will issue a certificate for that gear type only.

For persons who have already qualified as Senior Operators with one gear type, a further 5 sessions will be the requirement to gain authorization for additional electrofishing types. Trainees will need to keep a separate log of their hours of use with each gear type.

First-aid certificates and medicals must be updated every two years.

Construction Standards of Equipment

General

1. All electrofishing equipment must be constructed only by qualified electrical mechanics and electronic technicians.

   Wiring and general construction must comply with the standards of country's regulations.

2. Each electrofishing unit must be accompanied by a detailed instruction manual and gear log book that contains hours and places of use, operator's names, maintenance and electrical checklists.

3. Mains electricity must **never** be used for any electrofishing operation. Non-standard plugs must be used to prevent connection to the public power supply.

4. The maximum voltage output allowable is 1000 V. Only DC current is to be used for the electrical fields due to its tendency to cause fewer injuries to fish and being less dangerous to operators in the event of an accident. Approval to use AC may be granted in special research circumstances only, *e.g.* for studies into the effects of different electrical fields on fish behavior, physiology and mortality.

5. All cable and connectors must be non-interchangeable, waterproof, and give adequate mechanical protection, and all cabling must be of multicore conductors.

6. Batteries must be sealed type or housed in spill proof marine battery containers.

7. Generator output must not be earthed but must be isolated from the frame so that power can only enter the water via electrodes. This involves removing the earth wire that grounds the frame to the output circuit. The frame however must still be earthed via an earth stake for shore mounted units, or by grounding to all other metal objects within a boat, whether the boat is metal or non-metal. In non-metal boats, the outboard motor must be grounded as well. Electrofishing generators must be clearly labelled and must not be used for other purposes.

8. Generators and control boxes must not be carried while running or energized (except specifically designed backpack units).

9. Control boxes must be IP34 rated, protected by current and/or thermal overload, and fitted with a large red latching "STOP" button to interrupt power supply from the generator to the control circuitry. An ammeter must be fitted to measure control box output, and a lamp to indicate when the unit is energized and the electrodes are live. It is recommended that a voltmeter and audible signal emitter be fitted as well. Large, easily visible warning signs indicating the danger of high voltage are to be attached.

Due to the cooling vents present on the control boxes of some electrofishing equipment, care must be taken to prevent the entry of water into the electronics when cleaning or during rain. A waterproof cover is recommended when the apparatus is not in use.

10. Power to the electrodes must be switched by at least one "deadman" or fail-safe switch. All control-switch circuits must be extra low voltage (< 32 V AC or < 115 V DC)- never directly switching the anode supply.

11. Hand held anodes must not be used in metal boats or be fitted with dip nets except when connected to a backpack rather than generator powered unit, although this procedure is not recommended. All anode poles and dip net handles must be made of a non-conductive material with no metal core but not wood. Carbon fiber reinforced poles are likely to be conductors of electricity and must be thoroughly tested for insulation properties if being considered for use.

Backpacks

1. Battery powered backpack units must use only fully sealed dry cells as a power source. Specifically designed generator powered units are not recommended, but if used it is recommended that they be fitted with earth leakage circuit breakers at the generator output.

2. The backpack unit must incorporate a quick release harness, a dead man switch on the anode pole, and be fitted with a mercury tilt switch that cuts off power from the battery or generator whenever the unit is tilted at more than a 45 degree angle. The tilt switch can have an automatic

Figure 15: Operating Back-Pack Electrofishing System in a Wadable Stream.

reset- although a manual button that can be reset by the operator is recommended. The unit must have an audible alarm when in use.

3. Backpack electrofishing is not recommended in water deeper than operator crotch depth.

4. If a backpack unit is operated from a boat, it must be ensured that the cathode is isolated from the boat hull. It is also recommended that two foot-pedal safety switches connected in series be incorporated.

5. Backpack units must meet IP 37 standards

Shore-Based Units

1. Generators must be safely anchored and not capable of falling or being pulled into the water by the anode cable. The generator must be manned at all times during electrofishing operations, and electrode and dip net crew must be in line of sight or in radio contact with the generator attendant.

2. Only one anode is permissible per generator unit. Generators and control boxes must not be carried while energized. Hand held anodes must incorporate a dead man switch on the anode pole.

3. Electrode and dip net crews may operate from a boat using a shore based generator in safe conditions providing the anode is fixed not hand held, and floating cable is recommended to minimize the risk of snagging. Boat size and requirements such as, foot switches and hand rails are as for boat-mounted units.

Boat-Mounted Units (min. 3.5 m)

1. The boat driver must be a holder of the relevant local boating licence and be sufficiently experienced with the size of boat being used and in the river or lake conditions present. The boat must be adequately stable and have ample free board when fully loaded with gear, crew, plus with a full catch of fish on commercial electrofishing boats.

2. For boats under 4.0m x 1.2m, a crew of only two is allowed and maximum generator size is 5kVA.

3. Anodes must be fitted to the bow and not capable of touching any part of the boat, and cables are to be channeled or clipped to boat sides to prevent tripping.

4. To reduce the risk of dipnetter staff accidents, fixed or removable hand rails of at least 70 cm high must be fitted and non-skid flooring is recommended.

5. The drivers of electrofishing boats must use foot operated dead man switches which must be operated simultaneously with boat netters, who must have either; (a) at least one foot operated switch (several may be connected in parallel if more than one netter is used), or (b) "life-line" belt cord cut-out switches.

6. All lighting and ancillary electrical equipment must be extra-low voltage (< 32 V AC or < 115 V DC).

7. Large red DANGER warning signs must be displayed on each side of the boat.

8. Generators and control boxes must be fixed in position during operation.

Suppliers

It is recommended that only approved, commercially produced equipment (*e.g.* Smith-Root, Coffelt, *etc.*) be used. Locally built units must undergo strict testing and comply with the checklist.

Maintenance Requirements

1. All equipment should be carefully inspected by a Senior Operator for mechanical faults, worn insulation, loose components and connections *etc.* prior to and following all operations.

2. Each electrofishing unit must have its own log book to record details of hours use, maintenance, repairs and inspection details.

3. Regular electrical safety checks must be performed by a licensed electrician on all electrofishing units, the minimum requirement being insulation and

voltage output testing. These checks must be performed every 12 months or 600 logged hours (whichever is reached first) for fisheries research agencies and universities, and every 6 months or 600 logged hours for commercial operations.

4. Multiple units (*e.g.* several backpacks) should be individually numbered to allow accurate records of maintenance and repairs to be kept.

Personal Insulation and Safety

1. Rubber boots or waders, plus 1000 V rated linesmen gloves must be worn by all electrofishing team members during operations.

2. Life jackets must be worn by backpacking and shore based crews (self-inflating jackets that use at least 33g carbon-di-oxide cylinders are permissible) in any dangerous situation where the water depth is greater than 50 cm. It is recommended that they or standard PFD type 1 or 2 jackets be worn by boat dip-netters as well.

Operations Practices

A minimum of two operators are required for all electrofishing except shore-based units where two plus the generator operator are needed.

Public Safety

1. Spectators must be warned to keep away, and no electrofishing can be performed within 50 m of other boats or shore viewers. In public areas, it is recommended that signs be erected warning people to keep their distance.

2. It is not recommended that observers be allowed on electrofishing boats during operations. If present, they must be warned of the dangers involved and fully equipped with all necessary safety equipment. They may be required to sign indemnity forms stating that they suffer from no major heart or respiratory complaint, and that the crew will not be held responsible in the event of an accident.

Weather and Site Limitations

Electrofishing should never be conducted in rain or rough water conditions. It is up to the Senior Operator to decide whether conditions, such as, current, wind, navigability and weather are safe and suitable or not.

Working Procedures

1. Senior operators are responsible for checking correct setup of equipment, crew safety, clothing and readiness before starting the generator. Anodes must be in the water, and are treated as live at any time the generator is running.

2. Relevant local authorities must be notified prior to any electrofishing, *e.g.* Fisheries Officers, National Parks and water management bodies.

3. Never put unprotected parts of the body in the water if the generator is running. Stunned fish must only be removed from the water using insulated dip nets – never attempt to grab a fish by hand !

4. If anodes become fouled on snags or trees, the generator must be stopped before attempting to free them.

5. No smoking is allowed in the vicinity of fuel tanks or during operations.

6. It is recommended that the electrofishing team have access to communications at all times in case of emergency.

7. All equipment should be stored in a dry, clean and secure place.

Care of Fish

Only the minimum power necessary to attract and stun the fish effectively should be used. Contact of fish with live anodes should be avoided, as the resulting shock will be much greater. If threatened species are observed that are not being targeted, appropriate measures must be taken to minimize disturbance and stress to these fishes. Commercial operations must also take action to minimize harm to any species that they are not authorized to catch under the conditions of their licence.

Care of other Fauna

1. Electrofishing must be halted within 15 m of any animals standing in or about to drink from the water, or in contact with a wire fence line that enters the water.

2. The utmost care possible should be taken to avoid shocking platypus, birds and other native aquatic animals.

Transfer of Biological Material

The utmost care must be taken to prevent the transfer of biological material between waterways. There are risks of introducing major weeds such as water hyacinth, noxious fish; or disease organisms such as EHN virus.

All gear (including nets, traps, boats, trailer *etc.*) must be thoroughly cleaned of plant material and sun dried. The live-well and deck of electrofishing boats should be flushed with pool chlorine solution or sodium metabisulphite when on land to ensure no material is transferred.

Extreme care must be taken when entering pristine or protected (such as municipal water supplies) waterways, or when sampling is done in waterways infested with weeds or alien species.

Accidents and Emergency Procedures

Appropriate first aid kits and dry chemical type fir extinguishers should be present on all electrofishing boats larger than 4.5 m, in the vicinity of all other gear types, and be readily accessible in an emergency.

In the event of accident;

1. Switch off power supply at control box;
2. Stop generator;
3. If casualty is consciousness, check pulse and respiration;
4. Begin resuscitation if either is absent;
5. Apply first-aid for any injuries;
6. Have any electric-shock casualty examined by a doctor as soon as possible even if they appear to be fully recovered;
7. Report all accidents involving electrocution to the National Electrofishing Safety and Training Administrator within 14 days.

Any generator, control box or backpack unit is not of a completely sealed type and is accidentally immersed in water must be removed from service immediately and tested for water damage by a licensed electrician before further use.

Electrofishing Apparatus Safety Features Checklists

Equipment Type – Shore-based

Equipment No. ——————————————————————

Manufacturer of equipment ——————————————————

Owner of equipment —————————————————————

Check performed by —————————————————————

Date—————————————————— Log hours—————————

Date of last inspection ———————————— Log hours ————

Generator	*Yes/No*
Earth link disconnected	————
Labelled "For electrofishing only"	————
Output socket only compatible with control box cable	————
Moving parts and exhaust adequately guarded	————

Control box	
IP 34 rated	————
Output voltage maximum 1000 V	————
DC output only (commercial operations)	————
DC output available (research operations)	————
All wiring in standard multi-core	————
Ammeter measuring control box output	————
Indicator light	————
"Warning High Voltage" signs present	————

Double pole latching " **STOP**" button on input   ——————

Input plug only compatible with electric fishing generator   ——————

Anode

Submersible 3 "dead man" control switch on anode pole   ——————

Extra-low voltage control circuit   ——————

Anode pole made from non-conductive material   ——————

Equipment Type – Boat-mounted

Equipment No. ————————————————————————————

Manufacturer of equipment ——————————————————————

Owner of equipment ——————————————————————————

Check performed by ——————————————————————————

Date————————————————————Log hours—————————

Date of last inspection ————————————Log hours —————————

Generator

Earth link disconnected   ——————

Labelled "For electro-fishing only"   ——————

Output socket only compatible with control box cables   ——————

Moving parts and exhaust adequately guarded   ——————

Control box

IP 34 rated   ——————

Output voltage maximum 1000 V   ——————

DC output only (commercial operations)   ——————

DC output available (research operations)   ——————

All wiring in standard multi-core   ——————

Ammeter measuring control box output   ——————

Indicator light   ——————

"**Warning- High Voltage**" signs present   ——————

Double pole latching "STOP" button on input   ——————

Input plug only compatible with electrofishing generator   ——————

Anodes

Extra-low voltage control circuit   ——————

Anode poles made from non-conductive material   ——————

Boat

Submersible "dead man" foot switch for driver and netter ————————

Electrical cables channeled or clipped to boat sides ————————

Hand rails for dip netters at hip height ————————

All lighting and ancillary equipment extra-low voltage ————————

Equipment Type – Back-pack

Equipment No. ————————————————————————

Manufacturer of equipment ——————————————————————

Owner of equipment ————————————————————————

Check performed by ————————————————————————

Date————————————————————Log hours————————

Date of last inspection ————————————Log hours ————————

Back-pack

IP 37 rated ————————

Output voltage maximum 1000 V ————————

DC output only (commercial operations) ————————

DC output available (research operations) ————————

If battery powered- only sealed dry cell batteries ————————

Quick release harness ————————

Anode

Submersible " dead man " control switch on anode pole ————————

Anode pole made from non-conductive material ————————

Chapter 14

Electrofishing Terminology*

General

Alternating current: cyclic current, the mean value of which is nil during a total period. An alternating current is characterized by a sequence of positive and negative waves that are equal, usually sinusoidal, and follow each other alternatively at regular intervals.

Anode: the positive electrode, usually hung from a boom extending away from the electrofishing vessel.

Aperiodic impulses: impulses following each other at varying time intervals.

Applied power density: power density available for transfer to a fish at a particular location in the water.

Bonding: the permanent joining of metallic parts to from an electrically conductive path that ensures electrical continuity, with the capacity to safely conduct current.

Branch circuit: the circuit conductors between the final overcurrent device protecting the circuit and the electrical load(s).

Cathode: the negative electrode, usually located on the hull of the electrofishing vessel.

Circuit breakers: a device designed to open and close a circuit by a non-automatic means, and to open the circuit automatically on the predetermined overcurrent without damage to itself when properly applied within its rating.

Complex pulse system: a complex pulse train or burst form of pulsed direct current by Coffelt Manufacturing, Inc., developed in response to recently detected high mortality rates caused by commercially available PDC wave forms.

Condenser discharges: current composed of a steady sequence of exponential discharges.

Conductivity: the ratio of the density of the unvarying current in a conductor to the voltage gradient that produces it; the common unit of measurement is the μsiemen/cm = μmhos/cm.

Conductance: the measure of the ability of a component to conduct electricity, the reciprocal of resistance; the unit of measurement is the siemen (mho).

Current: the rate of electrical charge flow in a circuit; the practical unit is the ampere (amps), which is one coulomb per second.

Current shape: the geometric shape of the current during cycle; usually this refers to the rate of growth and decay of an impulse.

Cycle: one full revolution of a periodic phenomenon.

Deadman switch: a switch that requires constant pressure to supply electrical current to the circuit.

Direct current (continuous, galvanic): unidirectional constant current.

Effective fish conductivity: the apparent conductivity of live fish as determined by statistically fitting electroshock response data to the theoretical curve developed for the concept of constant power.

Electrical charge: a fundamental property of matter that can be classified as a fundamental physical quantity; the practical unit is the coulomb. The electron, the smallest charge identified in nature, has a magnitude of 1.6×10^{-19} coulomb.

Electrofishing: the use of electricity to provide a sufficient electrical stimulus in fish to permit easy capture by netting.

Frequency: total number of cycles per unit time measured in hertz; 1 Hz equals one cycle per second.

Ground: a conducting connection between an electrical circuit or equipment and the earth, or to some conducting body that serves in place of the earth.

Half-wave rectified current: current composed of a sequence of half sine waves in the same direction, separated by pauses of equal duration. This is obtained by passing an alternating current through a rectifier.

Impulse: electric phenomenon of short duration compared with the period.

Interrupted direct current: unidirectional current interrupted by periodic pauses.

Isolation transformer: a transformer inserted into a system to separate one section of the system from undesired influences with other sections.

Mismatch ratio: the ratio of either the two resistance values or two conductivity values determined for adjoining media. For electrofishing, this is the ratio of conductivity of the water to the effective conductivity of the fish.

Multiphase current: the number of phases, the whole of n being alternating currents originating from the same source and out of phase with each other by 1/n of a period (1/3 of a period with three phase current).

Netter: the individual who nets the captures fish during electrofishing operations.

Pause duration: interval between two electric phenomena.

Period: time interval between two identical stages in an electric sequence.

Power: the rate of doing work or the energy-per-unit-of-time; the practical unit is the watt (W), which is one joule per second.

Applied power: incidental power at an electrical interface separating two mediae.

Constant transferred power: the constant value of transferred power desired under all conditions of mismatch.

Maximum output power: the maximum available power delivered to an external load from a power source having an internal resistance equal to that of the external load.

Reflected power: the portion of applied power that is not transferred to the second medium.

Transferred power: the portion of applied power transferred from the first medium to second medium.

Power control circuit: the circuit that interconnects and adjusts the power from the pulsator or generator to the electrodes.

Power density: the power or energy-per-unit-of-time dissipated in a given volume of material. The unit measurement is watts per cubic centimeter (W/cm3).

Power density in fish: the desired constant value of power density to be transferred to a fish. Also, the threshold of in vivo power density required to produce a specific electroshock response at a specific conductivity.

Pulse duration (pulse length): duration of an impulse.

Pulsed current (pulsating AC or DC): unidirectional current composed of a sequence of cyclic impulses.

Quarter-sine wave current: a special kind if current electronically obtained from alternating current from V-max to zero.

Rectified alternating current: current composed of an uninterrupted sequence of half sine waves in the same direction, and obtained from an alternating current by means of a four-way bridge rectifier. Also called, full-wave rectified current.

Resistance: the ability to react to the flow of AC or DC with an opposition to the flow of current. Also, the ratio of the applied voltage to the induced current that it produces; the unit of measurement is the ohm.

Resistivity: the reciprocal of conductivity; the common unit of measurement is the ohm-cm.

Smooth rectified current: direct current derived from alternating current by using rectifiers and a suitable capacitance inductance filter. When insufficiently filtered, the current shows weak sinusoidal variations and is called ' partly smoothed rectified current', or ripple current, or undulating current.

Square wave (syn. Rectangular pulses): cyclic waveform with steep rise and fall time, with flat top and bottom.

Variable voltage pulsator electroshocker: the device used to deliver the pulsed electric current.

Volts or Voltage: the energy-per-unit-of-electrical-charge; the volt (V) is the unit of measure where one volt is one joule per coulomb.

Voltage gradient: the rate of change of voltage with distance. Also, the force-per-unit-of-electrical-charge; the common unit of measurement is volts per centimeter (V/cm).

Electric Field Characteristics

Anode (or cathode) field: in electric fishing, field around the electrode beyond which the values of potential gradient are unimportant.

Conductivity (of water): conductance of 1 cubic cm. of water. Conductivity is the inverse of resistivity.

Critical zone of current density: in electric fishing, current density area around the electrode in which a fish is shocked.

Current lines (flow lines, equiflux): imaginary lines that represent direction of current flow perpendicular to equipotential surfaces.

Density (of current): current density passing through one unit of cross-sectional area perpendicular to the current lines of an electric field.

Equipotential surface: a surface on which all points are at the same electrical potential. Equipotential surfaces are perpendicular to the direction of current flow.

External resistance: electrical resistance between electrodes.

Heterogeneous field: field in which current density and potential gradient decrease as a function of the distance from electrodes.

Horizontal field: see vertical field

Isolines: lines of equal potential gradient.

Moving field (syn. Movable field): field in which surfaces of equal relative potential (related to the supply voltage) are displaced as a function of time (rotating field, intersecting field, *etc.*).

Potential gradient: potential difference in an electric field-per-unit-length on the direction perpendicular to the equipotential surfaces; this gradient is measured in volts per centimeter (V/cm).

Resistivity (of water): resistance of 1 cubic cm. Resistivity is the inverse of conductivity.

Stationary field: field in which surfaces of equal potential (related to the supply voltage) are steady.

Vertical field: field in which the potential gradient is lower on a ground plane than on a vertical plane, so that a fish swimming horizontally into the field will be subject to a body voltage much lower than if the field were horizontal itself at the same distance from the electrode.

Behavior and Physiology

Anelectrotonus: decrease of nerve excitability on the anode side.

Anodic (cathodic) curvature: curving of the fish body towards the anode (cathode) under the influence of a unidirectional current, when the fish is perpendicular to the current lines.

Ascending current: according to conventional direction of current (from + towards -), electric current ascending into the system from the periphery towards the fish nervous centers, occurring when the fish is facing the cathode.

Autorhythm: excitability of nerve and muscle provoked and sustained by a constant continuous current.

Body voltage: measured potential difference between head and tail of a fish in an electric field.

Catelectrotonus: increase of nerve excitability on the cathode side of a shocked fish.

Closing of the circuit reaction: nerve or muscle excitation produced by closing the circuit.

Descending current: electric current going down into the system from nervous centers twards the periphery (see ascending current), as in the case of fish facing the anode.

Electrotaxis: fish swimming induced by any kind of electric current.

Fixation: state of immobility of fish resulting from tetanus under the action of electric current, distinct from galvanonarcosis.

Forced swimming (first swimming towards the anode) a very fast swimming motion towards the anode, induced by a constant current.

Frightening effect: fish escape from an electrode under the action of current.

Galvanonarcosis: state of immobility of fish resulting from muscular slackening under pulsating direct current.

Inhibition of swimming: slowing down of swimming movements; produced by a low and constant continuous current when a fish faces the anode.

Narcosis: state of immobility resulting from muscular slackening.

Opening of the circuit reaction: nerve or muscle excitation produced by opening the circuit.

Oscillotaxis: swimming artificially induced by an alternating current.

Pseudo-forced swimming (second swimming towards the anode): out of balance swimming produced by a strong and constant continuous current. Occurs when a fish faces the anode.

Rheobase: minimal intensity of current indefinitely maintained to release the excitation of nerve of muscle.

Spatial summation: cumulative effect produced on a neuron by means of several simultaneous stimuli.

Taxis: artificial swimming induced by a stimulating agent.

Temporal summation: cumulative effect produced on a neuron by a series of stimuli.

Tetanus: state of muscular rigidity.

Threshold: minimal value of current parameter inducing a determined reaction.

Useful time: minimal time during which an electric current of a given value must be maintained to produce an excitation.

Source

Adapted from Cowx, I. G. and P. Lamarque, 1990. Fishing with Electricity, Application in Freshwater Fisheries Management; and Koltz, A. K. 1989. A Power Transfer Theory for Electrofishing.

References

Allen-Gil, S.M. New Perspectives in Electrofishing, EPA/600/R-99/108, 2000.

Alexis E. Knight Electrofishing construction practices, US, Fish and Wildl.Ser. New Hampshire, 03246, 1978.

Alberta Govt. Electrofishing policy respecting injuries to fish, Fisheries Management Branch, 2012.

Australian Govt. Australian Code of Electrofishing Practice, NSW Fisheries Management Publication No. 1, 1997.

Adams, W.J, D. J Behmer, and W. O. Weingarten. Recovery of shocked common shiner, *Notropis conutus*, related to electric energy, Trans. Am. Fish. Soc. 101(3), 1972.

Amiro, P. G. Developments in Electric Fishing, Fishing News Books, Blackwell Scientific Publications, Ltd. 1990.

Armstrong, M.C. and J. H. Mundic. Floating fish shocker, Prog. Fish. Cult. 45(4), 1983.

Bernstein, Theodore Design considerations for an electrofishing power supply, Proc. IEEE Southeast Conf. Cal. No. 73, CHO 785-2 REG III, 1973.

Biswas, K. P. Preliminary observations on the effect of electric seine on fish Catch, Fishery Technology, Vol. VIII, No. 2, 1970.

Biswas, K. P. Electrophysiology and Fish Behavior, Daya Publishing House Astral International Pvt. Ltd., New Delhi, 110002, 2015.

Biswas, K. P. and N. A. Talwar. On the Development of an Electric Fish Catcher (Per. Commn).

Balayave, L. A. The behavior of ecologically different fish in electric fields, II Threshold of anode reactions and tetanus, J. Ichthy. 21, 1981.

Bird, D and I. G. Cowx. Developments in Electric Fishing, Oxford, Fishing News Books Blackwell Scientific Publications Ltd., 1990.

Bowles, F.J., A.A. Frake and R.H.K. Mann. A comparison of efficiency between two electric fishing techniques on a section of the river, Avon, Hampshire in I.G. Cowx ed. Developments in Electric Fishing, Oxford, Fishing News Books, Blackwell Scientific Publications Ltd., 1990.

Cowx, I.G. and P. Lamarque. Fishing with Electricity, Application in Freshwater Fisheries Management, Fishing News Books, 1990.

Chmielewski, A. Study of reactions and behavior of fish in heterogeneous field with single and multi-phase current, Fish Screen and Guides, 1964.

Cowx, I.G., A. Wheatly and P. Hickley. Development of boom electric fishing equipment for use in large rivers and canals in the United Kingdom, Aquacult. Fish. Manage, 19, 1988.

Cowx, I. G. Developments in Electric Fishing, Oxford: Fishing News Books, Blackwell Scientific Publications Ltd., 1990.

Cross, O.G. and B. Stott. The effect of electric fishing on subsequent captures of fish. Journal of Fish Biology, 7, 1975.

Dolan, C.R. and L.E. Miranda Immobilization thresholds of electrofishing relative to fish size transaction of the American Fisheries Society, 132, 2003.

Darrel, E. Snyder Conclusions from a review of electrofishing and its harmful effects on fish, Fish Biology and Fisheries, 13, 2003.

Edwards, J.L. and J. D. Higgins. The effects of electric current on fish. Final Technical Report Projects B-397, B-400 and E-301, Game and Fish Division, Department of Natural Resources, Atlanta, Georgia, 1973.

Eloranta, A. Electric fishing in stony Littoral Zone of Lakes, in I.G. Cowx ed. Developments in Electric Fishing, Oxford: Fishing News Books Blackwell Scientific Publications Ltd., 1990.

Eloranta, A., E. Jutila and S. Kanno. Electrofishing and its safety requirements in I. G. Cowx ed. Developments in Electric Fishing, Oxford: Fishing News Books, Blackwell Scientific Publications Ltd., 1990.

Funk, J. L. Wider application of electrical method of collecting fish, Trans. American Fisheries Society.

Frank M. Panek and Christine, L.D. Electrofishing and the effects of depletion, Sampling on Fish Health, A review and recommendations for additional syudy, US Geological Survey, National Fish Health Research Laboratory, West Virginia, 25430, USA.

Frankenberger, L. Application of a boat rigged direct current shocker on lakes and streams in West Central Wisconsin, Prog. Fish. Cult. 22, 1960.

Fredenberg, W. Evaluation of electrofishing induced spinal injuries resulting from field electrofishing surveys in Montana Dept. Fish. Wildl. Parks, 1002.

Godfrey, H. Mortalities among developing trout and salmon ova following shock by direct current electrical fishing gear, J. Fish. Res. Bd. Canada, 14(2), 1952.

Godfrey, H. Catches of fish in New Brunswick streams by direct current electrofishing, The Canadian Fish Culturist, No, 19, 1956.

Hosl, A. Dangers and precautions in the electrical fishery, Electro-Beratung, Bayern, Munich, Germany.

Hauck, F.R. Some harmful effects of the electroshocker on large rainbow trout, Trans. Am. Fish. Soc., 77, 1949.

Hickley, P. Aspects of fishing electrode design, Aquaculture and Fisheries Management, 1, 1985.

Hickley, P. and A. Starkie. Cost-effective sampling of fish populations in large bodies of water. J. Fish. Biol. 27 (Supplement A), 1985.

Hickley, P. and B. Millwood. The United Kingdom safety guidelines for electric fishing: its relevance and application, in I.G. Cowx ed, Developments in Electric Fishing, Oxford: Fishing News Books, Blackwell Scientific Publications Ltd., 1990.

Hollender, B Injury of wild brook trout by backpack electrofishing, page 13 in Western Division American Fisheries Society, Colorado State University Program, 1992.

Holmes, R., D. Mc Bride, T. Vivant and J.B. Reynolds. Electrofishing mortality and injury to rainbow trout, Artic grayling, humpback whitefish, least cisco and northern pike, Fishery Manuscript 90-3, Alaska Dept. Fish and Game Anchorage, 1990

Horak, D.L. and W.D. Klien. Influence of capture methods on fishing success, stamina and mortality of rainbow trout in Colorado, Trans. Am. Fish Soc., 96, 1967.

Hudy, M. Rainbow trout and brook trout mortality from high voltage AC Electrofishing in a controlled environment, N. Am. J. Fish. Manage. 5, 1985.

Hudy, M. Mortality from high voltage AC electrofishing, N. Am. J. Fish., Manage. 6, 1986.

Jesien, R and P. R. Hocutt. Method for evaluating fish response to electric fields, P 10-12, in I.G. Cowx ed. Developments in Electric Fishing, Oxford, Fishing News Books, Blackwell Scientific Publications Ltd., 1990.

Lamarque, P. Twenty years of electric fishing expeditions throught out the world, P. 344-531 in I.G. Cowx ed. Developments of Electric Fishing, Oxford: Fishing News Books, Blackwell Scientific Publications Ltd., 1990.

Latta, W.L. and G.F. Meyers. Night use of direct current electric shocker to collect trout in lakes, Trans. Am. Fish. Soc., 90, 81-83, 1961.

Lazauski, H.G. and S.P. Malvestuto. Electric fishing; results of a survey on boat construction, configuration and safety in the United States, P 327-329 in I.G. Cowx ed. Developments in Electric Fishing, Oxford: Fishing News Books, Blackwell Scientific Publications Ltd., 1990.

Luiq, W. Daming, X. Ronngong and L. Jeifu. A method of improving fishing efficiency in lakes by using a seine net with pulsed current, P. 41-45 in I.G. Cowx ed., Developments in Electric Fishing, Oxford: Fishing News Books, Blackwell Scientific Publications Ltd., 1990.

Lui, Q. Development of the model SC-3 alternating current scan fish driving device, P. 46-50 in I.G. Cowx ed, Developments in Electric Fishing, Oxford: Fishing News Books, Blackwell Scientific Publications Ltd., 1990.

Ludemann, K. and Koschinski, S. Sustainable brown shrimp fishery- is pulse fishing a promising http/pulse fishing/en/sites/pulse fishing files/pf research paper/ WWF-Technical Report, Sustainable-Brown-Shrimp-Fishery-2014 pdf.

Meyer-Waarden, P.F. Electrical Fishing, FAO Fisheries Study No. 7. Food and Agriculture Organization of the United Nations, Rome, 1957.

Mahoney, B.D., T.K. Iverson and S.B. Mathews. Synopsis and annotated bibliography on electrofishing with special reference to Columbia river squawfish control, Fisheries Research Institute, School of Fisheries, University of Washington, Seattle, Washington 98195, 1993.

Mc Iain, A.L. The control of upstream movement of fish with pulsed direct current, US Deptt. Of Interior, Fish and Wildlife Service, 1956.

Mann, R.H.K. and T. Penczak. The efficiency of a new electrofishing technique in determining fish numbers in a large river in Central Poland, J. Fish. Biol. 24, 173-185, 1984.

Malvestuto, S.P. and B.J. Sonski. Catch rate and stock structure: a comparison of daytime versus nighttime electric fishing on West Point Reservoir, Georgia, P. 210-218 in I.G. Cowx ed, Developments in Electric Fishing, Oxford: Fishing News Books, Blackwell Scientific Publications Ltd., 1990.

Mc Crimmon, H.R. and B. Bidgood. Abnormal vertebrae in the rainbow trout with particular reference to electrofishing, Trans. Am. Fish. Soc. 94: 84-88, 1965.

Mc Lean, I.A. Safety in electric fishing: a United Kingdom view, P. 324-326 In I.G. Cowx ed, Development in Electric Fishing, Oxford: Fishing News Books, Blackwell Scientific Publications Ltd. 1990.

Norman, G.S. and Jane Sharber. Epilepsy as a unifying principle in electrofishing theory, Trans. Am. Fish. Soc., 128, 666-671, 1999.

Newman, H.W. Effect of field polarity in guiding salmon fingerlings by electricity, SSR No. 319, US Fish. Wildl. Serv. Washington DC, 1959.

Nielsen, J.L. Electrofishing California's Endangered Fish Populations, Fisheries Management, Vol. 23, No. 12, 1998.

Newman, L.E. Spinal injury of walleye caused by PDC electrofishing, p. 14 in Western Division of the American Fisheries Society, Maryland, 1992.

Novotny, D.W. and G.R. Priegel. Electrofishing boats, improved designs and operational guidelines to increase the effectiveness of boom shockers, Technical Bulletin No. 73, Deptt. Of Nat. Resources, Madison Wisconsin, 1974.

Patten, B.G. and C.C. Gillaspie. The Breau of Commercial Fisheries Type IV Electrofishing Shocker, its characteristics and operation, SSR, US Fish and Wildlife Service, Washington DC, 1966.

Pratt, V.S. A measure of the efficiency of alternating current and direct Current Fish Shockers, Idaho Cooperative Wildlife Research Unit, University of Idaho, Mosco, Trans. Am. Fish. Soc., Vol 81, 1951.

Pratt, V. S. Fish mortality caused by electrical shockers, Trans. Am. Fish. Soc., 1957.

Paragamian, V.L. A comparison of day and night electrofishing, size-structure and catch-per-unit-effort for small mouth bass, N. AM. J. Fish. Manage. 9: 500-503, 1989.

Penczack, T. and H. Jakubowski. Drawbacks of electric fishing in rivers, p. 115-122 in I.G. Cowx ed., Developments in Electric Fishing, Oxford:Fishing News Books, Blackwell Scientific Publications Ltd., 1990.

Pierce, R.B., D.N. Coble and S.D. Corley. Influence of river stage on shoreline electrofishing catches in the Upper Mississippi River, Trans. Am. Fish. Soc. 114: 857-860, 1985.

Pratt, V. S. Fish mortality caused by electrical shockers, Trans. Am.Fish. Soc. 84: 93-96, 1954.

Pugh, J.R., G.E. Monan and J.R. Smith. Effects of water velocity on the fish guiding efficiency of an electrical guiding system, US Fish. Wildl. Serv. Fish Bull. 68(2), 307-324, 1970.

Quirijns, F., Strietman, W.J., Marlen, B.Van and Rasenberg, M. Flat fish pulse fishing research results and knowledge gaps//Retrieved from htpp//pulse fishing eu/Sites/pulse fishingen/files/pf research paper/C091. 15 Rapport Flatfish pulse Fishing gaps//-ss-ics pdf.

Rasenberg, M., van Overjee, H, Quirijns, F., Warmerdam, M., Van Os, B, and Rink, G. Monitoring catches in the pulse fishery Report No. C112/13 (P. 59)., 2013.

Randall, R.G. Effect of water temperature, depth, conductivity and survey area on the catchability of juvenile Atlantic salmon by electric fishing in New Brunswick streams, P. 79-80 in I.G. Cowx ed, Development in Electric Fishing, Oxford: Fishing News Books, Blackwell Scientific Publications Ltd., 1990.

Reynolds, J.B., S.M. Roach and T.T. Taube. Injury and survival of northern pike and rainbow trout captured by by electrofishing, P. 15 in Western Division of American Fisheries Maryland, 1992.

Stewart, P. and Stewart, L. An investigation into the effects of electric fishing equipment on Salmon and Sea Trout within the area of the Lancashire River Board, WYE River Board, 1960.

Shetter, D.S. The electric shocker and its use in Michigan Streams, Progressive Canadian Fish Culturist, Vol. 16, No. 9, 1947.

Simth, G.M.F. and P.F. Elson Direct current electrical fishing apparatus, Report presented to Fisheries Research Board of Canada, Fish. Cult. No. 9, 1950.

Saltveit, S.J. Studies on juvenile fish in large rivers, P. 109-114 in I.G. Cowx ed. Developments in Electric Fishing, Oxford: Fishing News Books, Blackwell Scientific Publications Ltd., 1990.

Sharber, N.G. and S.W. Carothers. Submerged electrically shielded live tank for electrofishing boats, N. Am. J. Fish. Manage. 7: 453-455, 1987.

Sharber, N.G. and S.W. Carothers. Influence of electric fishing pulse shape on spinal injuries in adult rainbow trout, P. 19-26 in I.G. Cowx ed, Development in Electric Fishing, Oxford: Fishing News Books, Blackwell Scientific Publications Ltd., 1990.

Simpson, D.E. and J.B. Reynolds. Use of boat-mounted electrofishing gear by fishery biologists in the United States, Prog. Fish. Cult. 3(2), 88-89, 1977.

Spencer, S.L. Internal injuries of largemouth bass and bluegill caused by electricity, Prog. Fish. Cult. 29: 168-169, 1967.

Steinmets, B. Electric fishing, some remarks on its use, P. 1-4 in I.G. Cowx ed, Development in Electric Fishing, Oxford: Fishing News Books, Blackwell Scientific Publications Ltd., 1990.

Sternin, V.G., I.V. Nikonorov and Y.K. Bumeister. Electrofishing, theory and practice, Translated from Rissian by Israel, Program for Scientific Translations, Jerusalem, 1976.

Soetaert, M., Decostere, A., Polet, H. and Verchueren, B, Chiers, K. Electro-trawling a promising alternative fishing technique warranting further exploration, Fish and Fisheries, 16(1), 104-124, doi: 10.1111/faf. 12047.

Stewart, P.A.M. Catch selectivity by electrical fishing systems, J. Cons. Int. Explor. Mer. 36(2), 106-106, 1975a.

Stewart, P.A.M. Comparative fishing for *Nephrops norvegiens* using a beam trawl fitted with electric ticklers, Marine Research, 1975/1 (P10) Aberdeen, 1975b.

Teal, L.R., J. Depestele, B. O'Neil, J. Cracymaersch D. Denderen, D. Van, R. R. Parker and A.D. Rijnsdrop. Effects of beam and pulse trawling on the benthic ecosystem (pp 1-53), Retrieved from http//edepot Wur. Nl/308956.,2014.

Temple, Gabriel, M. and Todd. N. Pearsons. Backpack and Drift Boat Electrofishing Design Considerations and Sampling Protocols, Washington, Deptt. of Fish and Wildlife, 2000.

Vibert, R. Neurophysiology of electric fishing, Trans. Am. Fish. Soc. 92(3), 265-275, 1963.

Verschueren, B. Pulse fishing in practice; presentation at dialogue meeting July 2015, Retrieved from htpp//pulse fishing en/sotes/pulse fishing en/files/150702 Presentation Bart Verschueren-0pdf, 2015.

Woolmer, A., E. Maxwell and W. Lart. SIPF C0083 – Effects of electrofishing for *Ensis* spp, on benthic macrofauna, epifauna and fish species, Seafish Report SR 652 (P 57), Grimsby, 2011.

Warry, F.Y., P. Reich, J.S. Hindell, J. McKenzie and A. Pickworth.Using new electrofishing technology to amp-up fish sampling in estuarine habitats, Journal of Fish Biology, Vol. 82, Issue 4, 1119-1137, 2013.

Weisser, J.W. and G.T. Klar. Electric fishing for sea lampreys (*Petromyzon marinus*) in the Great Lakes, P. 59-64 in I.G. Cowx ed, Developments in Electric Fishing, Oxford: Fishing News Books, Blackwell Scientific Publications Ltd., 1990.

Welton, J.S., W.R.C. Beaumont, and R.H.K. Mann. The use of boom-mounted multi-anode electric fishing equipment for a survey of the fish stocks of the Hampshire Avon, P. 236-242 in I.G. Cowx ed, Developments in Electric Fishing, Oxford: Fishing News Books, Blackwell Scientific Publications Ltd., 1990.

Whaley, R.A., O.E. Maughan and P.H. Whiley. Lethality of electroshock to two freshwater fishes, Prog. Fi Fish. Cult. 40(4); 161-163, 1978.

Willemstad, J. The electrified trawl as an alternative type of fishing gear to eel traps, P. 70-78 in I.G. Cowx ed, Developments in Electric Fishing, Oxford: Fishing News Books, Blackwell Scientific Publications Ltd., 1990.

Witt, A.J. and R.S. Cambell. Refinements of equipment and procedures in electrofishing. Trans. Am. Fish. Soc. 88: 33-35, 1959.

Wilber, R.S. and Edward F. Kilma. *In situ* experiments with coastal pelagic fishes to establish design criteria for electrical fish harvesting systems, Contribution No. 249, Southeast Fisheries Center, Fishery Bulletin, Vol. 72 No. 3, 1974

Yundt, S. Changes in catchability related to multiple electroshock, Proc. of the Annual Conference, Western Association of Fish and Wildlife Agencies 63: 116-123, 1983.

Yu. C, Chen, Z., Chen, L. and He. P. The rise and fall of electrical beam trawling for shrimps in the East China Sea; Technology, fishery and conservation implications, 1592-1597, 2007.

Index

A

Accidents and emergency 163
Alignment 11
Alternating current 9, 62, 167
Anelectrotonus 171
Anode 167, 170
Anode array 67
Anodic curvature 171
Applied power density 167
Ascending current 171
Autorhythm 171
Auxiliary generator 52

B

Backpacks 159, 166
Basic electricity 7
Battery operated gear 130
Boat and mechanical component 51
Boat mounted unit 161, 166
Body voltage 171
Bonding 167
Branch circuit 167

C

Care of fish 163
Catelectrotonus 171
Cathode array 67, 167
Checklists 164
Chemical effects 101
Circuit breakers 167
Circuit principles 15
Closing of circuit reaction 171
Comparative trials 102
Condenser discharge 168
Conductivities 22, 23, 168
Conductivity of water 170
Crew training 157
Critical zone 170
Current lines 170
Current shape 168
Current through water 40
Current types 8, 10
Cycle 168
Cylindrical electrode 75

D

Daily check list 152

Dead man switch 168

Density of current 170 Descending current 171

Design-physiological basis 81

Direct current 9, 62, 168

E

Effect on benthic invertebrates 101

Effect on cod 100

Effect on fish 101

Efficiency 45, 137

Electric fish catcher 92, 93

Electrical characteristics 68, 75, 105

Electrical charge 168

Electrical field theory 26

Electrical fishing in sea 97

Electrical fishing machine 119

Electrical harvesting system 106

Electrical system 51, 64

Electrodes 27, 40, 54, 60, 61, 67, 74, 131, 133

Electrofishing 1, 20, 69, 79, 86, 89, 168

Electrofishing boat 47, 63, 73

Electrofishing gear 98, 128

Electrofishing results 76, 78

Electrofishing technique 39, 86

Electrophysiology 13

Electrotaxis 12

Environmental effects 100

Equipment 43, 81

Equipotential surface 170

External resistance 170

F

Field study 61, 72

Fish conductivity 168

Fish reaction and sensitivity 19, 84

Fish stunner 127

Fisheries management implication 103

Fixation 171

Forced swimming 171

Frequency 168

Frightening effect 171

G

Galvanonarcosis 11, 171

Galvanotaxis 11

Gasoline engine generator 130

Ground 168

H

Half-wave rectified current 168

Harmful effect 141

Hazards of electro fishing 151, 157

Heterogeneous field 170

History 35

Horizontal field 170

Hovercran trawl 98, 105

I

Impacts on reproduction 149

Impulse 168

Impulse generator 132

Inhibition of swimming 171

Injuries and mortalities 146

Interconnection system 53

Interrupted current 168

Isolation transformer 168

Isolines 170

M

Main generator 51

Maximum output power 169

Mechanical configuration 63

Meters 52

Mid-water trawling 116
Mismatch ratio 168
Moving field 171
Multiphase current 168

N

Narcosis 171
Netless fish harvesting 112
Netter 169
Non-target species 145

O

Opening of circuit reaction 171
Operation 44, 71, 162
Oscilloscope 23, 24, 25
Oscillotaxis 9, 171
Overload protection 52

P

Pause duration 169
Period 169
Physiological effects 143
Policy and code of practice 155
Potential gradient 170
Power control 124
Power control circuit 169
Power density 169
Power source 91, 121, 122, 169
Power transfer 21
Principles of electric fishing 5
Problem areas 48
Pseudo-forced swimming 11, 171
Pulse duration 169
Pulse frequency 12
Pulse generator 112
Pulse trawling 98, 105
Pulsed direct current 12, 62. 169
Pulser 66

Q

Quarter sine wave current 169

R

Rectified alternating current 169
Reflected power 169
Resistance 169
Resistivity of water 170
Response of fish 99
Response of shell fish 99
Rheobase 172

S

Safety 29, 53, 122, 162
Seine 41
Shocker 41, 83, 127
Shore based unit 160
Smooth rectified current 169
Square sine wave 170
Standard equipment 158
Standardization 31
Stationary field 170
System operation 121

T

Taxis 172
Temporal summation 172
Testing in field 31
Tetanus 9, 11, 12
Threshold 172
Timing circuit 126
Transfer of biological materials 163
Transferred power 169
Transformer 65, 123, 124
Trawl path mortality 101

U

Unsuccessful experiments 70

Useful time 172

V

Variable voltage pulsator 170

Vertical field 171

Voltage 170

Voltage control 53

Voltage gradient 170

W

Water conductivity 59

Wave forms 17

Wolf power plant 80

Working procedures 162